Gale

Gale

4|02

ATLAS OF THE EVOLVING EARTH

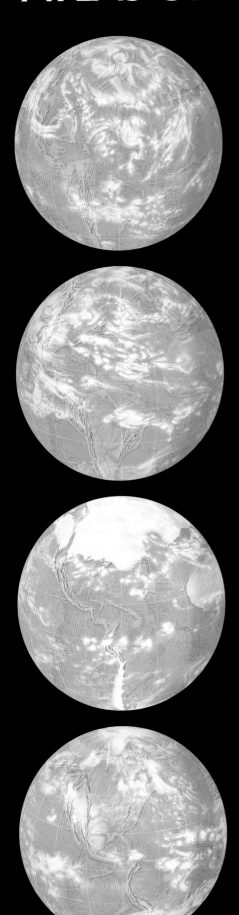

VOLUME

3

From the
Paleogene to the
Present

ATLAS
OF THE
EVOLVING
EARTH

VOLUME

3

From the
Paleogene to the
Present

IAN JENKINS

MACMILLAN REFERENCE USA
An imprint of the Gale Group

troit • New York • San Francisco • London • Boston • Woodbridge, CT

Academic consultant
Professor Michael J. Benton
University of Bristol, UK

Principal contributors
Dougal Dixon

Dr. Ian Jenkins
University of Bristol, UK

Professor Richard T. J. Moody
University of Kingston, UK

Dr. Andrey Yu. Zhuravlev
Paleontological Institute, Moscow

Project director Ayala Kingsley
Project editor Lauren Bourque
Art editors Ayala Kingsley, Martin Anderson
Cartographic manager Richard Watts
Cartographic editor Tim Williams
Paleogeography Dougal Dixon
Additional design Roger Hutchins
Picture research Alison Floyd
Picture management Claire Turner
Production director Clive Sparling
Proofreader Lynne Wycherley
Index Ann Barrett

Illustrators Julian and Janet Baker,
Robert and Rhoda Burns, Felicity Cole,
Dougal Dixon, Bill Donohoe, Brin Edwards,
Samantha Elmhurst, David Hardy,
Ron Hayward, Karen Hiscock, Ruth Lindsay,
Maltings Partnership, Denys Ovenden,
Colin Rose, David Russell, John Sibbick

Planned and produced by
Andromeda Oxford Limited
11–13 The Vineyard
Abingdon
Oxfordshire
OX14 3PX
England

www.andromeda.co.uk

Originated in South Africa by
Unifoto International, Cape Town

Printed in Hong Kong by H & Y Printing Ltd.

Published in the United States of America by
Macmillan Reference USA
1633 Broadway
New York,
NY 10019

ISBN 0-02-865632-6

Library of Congress Cataloging-in-Publication Data

Moody, Richard, 1939.
The atlas of the evolving earth / by Richard T. J. Moody...
p. cm.
Includes bibliographical references and index.
ISBN 0-02-865387-4 (hdb.)
 1. Physical geography--Maps for children. 2. Historical
geology--Maps for children. 3.Children's atlases. [1. Ph
geography. 2. Historical geology. 3. Atlases.] I. Title.

G1046.C1 .M6 2001
551.7--dc21

CONTENTS

VOLUME 3

INTRODUCTION

IN OUR MODERN AGE, most people are familiar with the broad outlines of the origin of the Earth, the rise of life in the sea, the age of the dinosaurs, early humans, and the ice ages. But it is astonishing that this story has been put together in only 200 years from scattered rocks exposed in quarries and on beaches, and from fossils picked up by chance.

Scientific observation of the natural world was practiced by the ancient Greeks and Romans, among others. In the year 1200, the Chinese naturalist and poet Zhu Xi wrote, "I have seen shells in the high mountains...the shells must have lived in water. The low places are now elevated high, and the soft material turned into hard stone." In Europe 600 years later, however, natural science had been coming along rather slowly. A few key points had been resolved, but the prevailing view was still of an Earth of very recent vintage, produced by a supreme creator.

Gradually, this view began to be challenged. In 1788 James Hutton, a Scottish landowner and amateur geologist, made a strong case for a hitherto unsuspected antiquity of truly stunning magnitude. In Scotland he oberved the processes of erosion and sediment deposition in the rivers and on the shores. He looked at ancient rock sequences, whose immense thickness suggested to him that they represented huge spans of time: "No vestige of a beginning, no prospect of an end." Hutton invented the principle of uniformitarianism, expressed in the axiom "The present is the key to the past," which means that the laws of nature are constant over time. With this, he laid to rest the medieval approach to geology and established it as a science.

In Hutton's lifetime, fossils moved from curiosities in private collections to the crux of the debate about the origins of life. Until 1750, most naturalists (who included many clergymen) had assumed that the Earth's plants and animals were as they had always been, and always would be: an extinction would mean that the Creator had made a serious mistake. However, with exploration and industrial digging, the remains of unknown plants and animals had begun to pile up.

Early explorers in North America sent their finds back to Europe for study. Shells and fern fronds did not pose a real problem, but around 1750 shipments of huge bones and teeth from surface deposits in the new territory of Ohio arrived in London and Paris. European scholars saw that these parts belonged to some kind of

elephant, but not the modern Indian or African elephant. Perhaps, they reasoned, some other elephant—they dubbed it *Incognitum* ("unknown")—still lived in the remote west of North America? This argument became impossible to sustain as explorers ventured further west without ever spotting a living elephant. By 1795, the great French anatomist and paleontologist Georges Cuvier announced that the American *Incognitum* was an extinct animal, the mastodon. He described other large beasts, known only from their fossilized bones, which were clearly extinct as well; they included Siberian mammoths and the giant South American ground sloth *Megatherium*.

Cuvier attributed the disappearance of these animals to global catastrophes that wiped out all life. This view was consistent with biblical tales of flood and plague, and was supported by traditionalists opposed to uniformitarianism, which argued for more gradual, steady change. As the science of geology developed, all evidence seemed to favor uniformitarianism, and until as recently as the 1960s most geologists were "ultra-uniformitarians," rejecting any

It was in the early 1800s that the intricate relationship of time, rocks, and fossils began to be deciphered.

process that could not still be observed in the present world. In fact, the catastrophists were right in many ways; mass extinctions may be attributed to events such as meteorite strikes and ice ages. However, even these are now known to be natural phenomena.

Evidence for the uniformitarian view came from the principles of stratigraphy—the ordering of rocks—which were largely worked out in the 1820s and 1830s. Hutton established a time frame for rocks; his successors noticed that particular distributions of rocks were repeated in many parts of the Earth. In addition, particular rock formations contained predictable sets of fossils. A rock unit from southern England could be correlated with another unit in Scotland or in France that had the same suite of fossils. Having correlated one unit, a geologist could predict what lay below and above. One by one, though not in chronological order, the key divisions of geological time—the Carboniferous, the Jurassic, the Cretaceous, the Silurian, and so on—were defined and named.

But what to make of the fossils? These had clearly changed over time. Did they represent a series of suc-

cessive creations and extinctions, as Georges Cuvier argued, or were they connected through the ages? Philosophers in England and France discussed this topic in the early part of the nineteenth century, but it was Charles Darwin who finally set out the principle and the mechanism in 1859. He showed that the diversity of life today could only have arisen by the splitting of evolutionary lines (speciations) over time, and that all life could be traced to common ancestors from unimaginably long ago. With this model, supported by more fossil finds, paleontologists in the nineteenth century drew a detailed picture of the history of life that has required little modification. Discoveries in the twentieth century shed further light, especially once the role of genetics was understood, but most contemporary work in paleontology consists of plugging gaps in the broad picture.

The discovery of a more modern animal such as a rabbit in 550-million-year-old rocks would upset the whole theory of evolution, but no such discovery has ever been made.

Geological science was revolutionized by two major advances around 1915. First came radiometric dating: the application of the principles of radioactive decay, discovered by Marie and Pierre Curie in the 1890s, to rocks. For the first time, geologists were able to establish absolute ages for rock sequences, and to put dates on the geological timescale that had been established in the 1830s.

The second revolution was the theory of continental drift. Until 1915, most geologists accepted that the Earth was stable. A few people had noted the jigsaw-puzzle fit between Africa and South America on maps, and others had noted similarities between fossils from widely separated locations. It was the German geologist Alfred Wegener who first insisted that none of this was coincidental. He argued that the continents had been part of the same giant landmass some 250 million years ago in the Permian and Triassic periods—and that the continents were still moving. Most geologists ridiculed his idea, quoting eminent geophysicists who declared that the Earth was solid and that there was no mechanism that could make the continents move.

Wegener's theory was only proven conclusively in the 1950s and early 1960s with discoveries in the deep oceans. The "motor" of continental drift is plate tectonics. Continents and oceans lie on separate plates that are pushed apart in the middle of the oceans by new crust welling up from the Earth's mantle. As the new rock emerges, the oceanic plates are pushed apart equally, but in opposite directions. In other places, to accommodate the new crust, plates dive down under each other, or push into each other. These movements raise mountains like the Andes and the Himalayas.

Geologists cannot study the events of every year throughout the history of the Earth, nor can paleontologists examine a fossil of every organism that has ever lived. But there is abundant evidence to reconstruct four and a half billion years of the Earth's history, and there are almost no unexpected or inconsistent discoveries. For example, thousands of observations show that much of North America and Europe was shrouded in ice one million years ago. No one has found million-year-old Canadian desert rocks or tropical reefs. Among the millions of new fossils uncovered each year, paleontologists have yet to be shocked by finding a rabbit in Cambrian shale, or a human among dinosaurs. Building on prior knowledge, it is possible to predict the gaps, and what might be discovered to fill them. This view might be said to be complacent. But until it is disproved by some extraordinary new find, rocks and fossils may be considered the keepers of the true history of the Earth and all its variety of life.

In this book you will read about the latest discoveries in geology and paleontology. The principles of stratigraphy, dating, and plate tectonics provide the framework, and detailed paleogeographic maps show the astonishing transformations of our planet. Supporting this is all the evidence accumulated by geologists of all nationalities as they piece together the habitats and climates of every corner of the world.

Volume 1 begins with the origin of the Earth, moving through the gradual changes that made it fit for life, up to the appearance of early life forms. Volume 2 continues the story with mountain-building as the continents shuffled and reshuffled, the development of forests, and a parade of animals from amphibians to dinosaurs and birds. Volume 3 presents the rapid transformations of more recent time, including the entrance of humans, and our unprecedented effect on the world. It's a story even more awesome than James Hutton could have imagined 200 years ago.

Michael Benton   University of Bristol, UK

THE GEOLOGICAL TIMESCALE

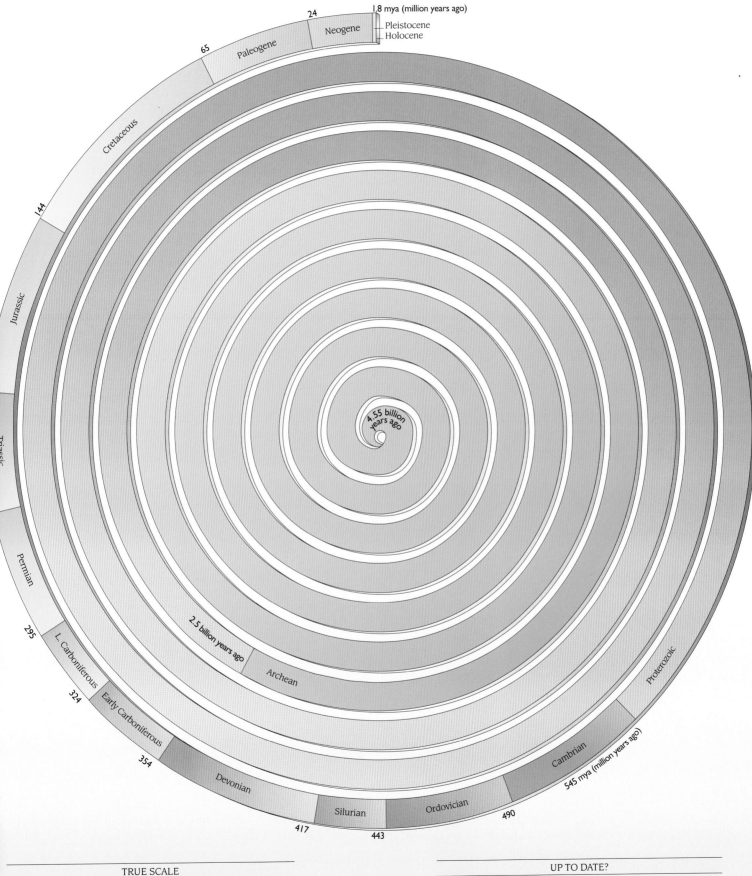

1.8 mya (million years ago)

24

Neogene

Pleistocene
Holocene

65

Paleogene

Cretaceous

144

Jurassic

Triassic

Permian

295

L. Carboniferous

324

Early Carboniferous

354

Devonian

417

Silurian

443

Ordovician

490

545 mya (million years ago)

Cambrian

Proterozoic

2.5 billion years ago

Archean

4.55 billion years ago

TRUE SCALE

This geological timescale has been designed as a spiral in order to offer a true linear measurement. It is therefore possible to compare the units of time from the Archean Eon, which lasted 2.05 billion years, to the recent Pleistocene Epoch, which lasted less than two million.

UP TO DATE?

Nothing in geology is more subject to change than dates, which are the focus of continual review. The data we have used throughout is largely based on the 1998 timetable by Bilal U. Haq and Frans W.B. van Eysinga (Elsevier Science BV) except where we have been otherwise advised.

THE
TERTIARY

65 – 1.8 MILLION YEARS AGO

WITH THE BEGINNING of the Cenozoic era 65 million years ago, the Earth entered the comparatively modern age. The transition was marked by the disappearance not only of the dinosaurs but also other species, such as ammonoids, the great marine reptiles, and the calcareous nannoplankton that had formed the extensive chalk deposits of previous ages. Unlike older strata, Cenozoic sediments are mostly soft (with the exception of some carbonates and siliciclastics from the earliest part of the Neogene); this makes them comparatively easy to recognize, and they are easily accessible, with plenty of fossils to distinguish them from the underlying Mesozoic. The traditional system divided the Cenozoic into two periods, the Tertiary and the Quaternary, the latter being the most recent. Modern geologists have found it more useful, however, to work with individual epochs within these two periods, which cover very short intervals of time.

THE NAME Tertiary is derived from eighteenth-century studies by European miners and budding geologists, who identified a tripartite division of the rocks in mainland Europe and the British isles. These scientific pioneers considered that the igneous and metamorphic rocks, which formed the basement of mountain ranges such as the Alps, had been the first to form on Earth during the chaotic days of creation. These "primary" rocks were overlain by "secondary" sedimentary rocks, which were predominantly fossil-bearing, and were thought to have been deposited during the Biblical Flood. The third type of rock, found in the low hills surrounding many European mountains, consisted of loosely-packed, stratified limestones, clays, and sands full of fossils that looked very much like living species. For this reason, these "tertiary" sediments were thought to have been deposited after the Flood. However, it was while examining the fossils of mammals in Tertiary rocks that the great French comparative anatomist and geologist Baron Georges Cuvier hit on the idea of extinctions. Cuvier realized that these mammals were significantly different to anything contemporary, and must have vanished from the world long ago.

The Tertiary, which encompasses the Paleogene and Neogene periods, began 65 million years ago at the Mesozoic–Cenozoic boundary and ended 1.8 million years ago as the world grew much cooler and drier, ending the warm, tropical conditions that had dominated the Mesozoic and the early Tertiary. The ice ages of the Quaternary (specifically, its earlier period, the Pleistocene) were the product of a mass of cold water, the psychrosphere, forming deep in the world's oceans during the Tertiary as Australia finally separated from Antarctica. Other major events during the Tertiary included the formation of the Panamanian Isthmus, which redirected warm water from the Atlantic towards northern Europe. Both the Alps and the Himalayas also rose during this time as a result of continental collisions.

ENVIRONMENTAL changes throughout the Tertiary led to a global radiation of mammals. Whales appeared in the newly cooled oceans, whose stock of small invertebrates organisms provided a rich food source for any animals that could feed on them in large enough quantities. On land, grasslands succeeded the tropical forests of the Mesozoic, providing an ecological niche into which large herbivorous mammals quickly expanded. Grass-eaters replaced many types of leaf-eaters, totally altering the composition of mammalian herbivore communities. Carnivores promptly appeared to feed upon the grass-eaters.

The Cenozoic Era has become known informally as "The Age of Mammals" due to the explosive diversification of this group as the dominant large-bodied animals. Mammals had first appeared during the late Triassic period of the Mesozoic, but there were few opportunities for them to expand into a world dominated by dinosaurs, which both competed for food (both flesh and plant) and preyed on smaller animals such as the

The Age of Mammals witnessed rapid and enormous diversification of a group of animals that had been unable to expand in the shadow of the dinosaurs.

early mammals. With the Cenozoic, all that changed. The small, mostly nocturnal, non-specialist mammals rapidly evolved into groups as diverse as whales, bats, and horses—nearly 100 families by the Eocene epoch. Elephants, cows, and cats and dogs made their appearance slightly later. Some of these animals were completely new; others had descended from lines that had existed in the Mesozoic. Not all of them survived into the succeeding Quaternary period; those that disappeared included long-legged ducks, giant carnivorous flightless birds, and *Indrichotherium* (formerly *Baluchitherium*), a member of the rhinoceros family and the largest mammal ever to have lived on land.

Among these expanding and diversifying mammals were the primates, which are characterized by their exceptional brain-power and social organization. The order began modestly enough, about 50 million years ago, with a tiny tree-dwelling animal that resembled a modern mouse lemur or tarsier; the larger specimens were only 2.2lbs (1kg) in weight. Their descendants colonized almost every part of the world during the later Paleogene and early Neogene. From such humble ancestors evolved the entire line of anthropoids, the group that eventually gave rise to humans. But this species was a much later arrival.

About 35 million years ago, in the early Oligocene, the straight-tailed, narrow-snouted "Old World" monkeys of what are now Africa and Eurasia appeared. The New World monkeys of South America, distinguished by their gripping tails and broad snouts, evolved later in the Oligocene from a group that had migrated out of Africa. Old World monkeys in turn gave rise to the apes, many of which became ground-dwelling. One ape family, the Ramapithecidae, arose in Africa about 17 million years ago and ranged in weight from 44lbs (20kg) to 600lbs (275kg). Among the diversity of apes—a far greater number than exist today—yet another new form appeared about 4 million years ago: the genus *Australopithecus*, an ape that walked upright. It was the oldest confirmed member of the human family,

THE
PALEOGENE
65 – 24
MILLION YEARS AGO

*T*HE PALEOGENE *is often overlooked in popular textbooks because it comes straight after the Cretaceous period and the demise of the dinosaurs. But much of what we see in our modern world had its origins in this important period of earth history. The most significant changes include the shift from the tropical forests of the Eocene to the cooler, open savannah of the Oligocene, and the accompanying reorganizations in the types of mammals that were abundant then.*

These global changes are linked to the formation of a   cold ocean current that circled Antarctica. As South America and Australia rifted from Antarctica, the warm current that had flowed along the continental margins was deflected by the arrival of a cold one, leading eventually to the formation of ice sheets. Today's Antarctic ice caps and the modern diversity of living mammals are the two great legacies of this portion of Earth history.

*T*HE GREAT geologist Charles Lyell's work formed an early basis for the current form of the "Tertiary" that comprises all the Cenozoic (the last 65 million years)

The dramatic disappearance of the dinosaurs at the end of the Cretaceous left many ecological niches vacant.

except for the Ice Ages that cover the past two million years, and for which the term Quaternary is used. Some experts subdivide the Cenozoic into two equal subdivisions: the Paleogene (65 to 24 million years ago) and the Neogene (24 million years ago to the present).

It is somewhat confusing that the most fundamental changes in Cenozoic history happened between the Middle Eocene and earliest Oligocene, and not at the formal boundary of any of the Cenozoic epochs. The establishment of the six-epoch Cenozoic period (including the three-epoch Paleogene), and continued refinement of the stratigraphy of these rocks, allowed

KEYWORDS

ALPINE OROGENY
—
ARCHAEOCETE
—
CARNIVORE
—
CONDYLARTH
—
CREODONT
—
HOT SPOT
—
LARAMIDE OROGENY
—
MAMMAL
—
MESONYCHID
—
RING OF FIRE
—
SEAFLOOR
SPREADING
—
TETHYS SEAWAY
—
UNGULATE

evolutionary changes that occurred throughout this part of Earth's history to be put into a clearer context.

The extinction of the dinosaurs at the boundary between the Cretaceous period and the Paleocene epoch changed the composition of the terrestrial vertebrate faunas completely. Ecological niches that had previously been filled by the dinosaurs were now available to be occupied by other animals. Roles such as those of large herbivores and large carnivores were taken up by the mammals. However, in the following epochs of the Paleogene, mammals defined many new ecological niches that had never before existed. Specializations such as those of anteater, grass grazer, and gnawer were originated by terrestrial mammals. Many of these bizarre-looking animals represent a response to what may be informally termed an evolutionary vacuum, in which ecological niches are made vacant by extinctions. However, during the early stages of this opportunistic refilling, the process of evolution produced a dazzling array of strange body plans and adaptations.

CRETACEOUS	65 mya	60		55		50	PALEOGENE
Series		PALEOCENE					EOCENE
European stages	DANIAN	SELANDIAN	THANETIAN	YPRESIAN			LUTETIAN
N. American stages	PUERCAN	TIFFANIAN		WASATCHIAN		BRIDGERIAN	
		└ TORREJONIAN		└ CLARKFORKIAN			
	Mid-Atlantic rift splits Greenland from N. America and Eurasia				Collision of Adriatic microplate begins the Alpine orogeny		
Geological events	Laramide orogeny continues; uplift of Rocky Mountains			Widespread development of new subduction zones in the P			
Climate			Warm, tropical to sub-tropical climate				
Sea level			Moderate to high, fluctuating				
Plant life			Tropical vegetation (lianas, cycads, etc) predominant				
Animal life	Adaptive radiation of mammals		• First "true" Carnivores			Creodonts are dominant carniv	
					└ First whales		

Mass extinction

This riot of mammalian evolution also extended into the marine realm, for it was during the Paleogene that the first whales appeared. Whales evolved from a group of Paleocene (early Paleogene) hoofed carnivores known as mesonychid condylarths. The earliest remains of whales come from Early–Middle Eocene rocks in Pakistan and Egypt. The seas of the Paleogene also witnessed other major events in the history of vertebrates, for it was then that modern sharks flourished. The cartilaginous skeletons of sharks soon disintegrate, but their hard teeth do not. Collecting fossil shark's teeth from Paleogene rocks has always been a popular pastime for amateur and professional alike.

In the middle of the seventeenth century the Danish physician Nicolas Steno, while dissecting the head of a large shark, noted similarities between its teeth and the curious "Tongue Stones" that had for centuries been dug out of soft rocks in cliffs on the island of Malta. In 1667 Steno published a book (*The Head of the Shark Dissected*) in which he argued that tongue stones were really the teeth of long-dead sharks that had become entombed during the biblical flood. His argument recognized that features of living creatures can be discovered in non-living objects now called fossils, and so Steno the physician became the world's first vertebrate paleontologist. However, the importance and historical legacies of the Paleogene rocks that gave Steno his material are recognized with increasing resolution today.

Major climatic changes occurred throughout the Paleogene Period. It was the start of a time of long-term cooling, although with numerous fluctuations. After an initial cool period during the Paleocene epoch, there was a return to tropical global conditions at the transition from the Paleocene to the Eocene. The collision of the Indian continental plate into the Asian continental plate in Late Paleocene/Early Eocene times (51 to 56 million years ago) is almost certainly linked to this—one of the warmest periods of the last 550 million years. This was followed by a significant alteration from the "greenhouse" global climate of the late Eocene to the icehouse conditions at the start of the Oligocene.

As THE Indian plate drifted north throughout much of the preceding Cretaceous period, volcanoes erupted in its path, waning as India collided with the vast Asian plate.

Volcanism and the release into the atmosphere of carbon from uplifted sediments contributed to global warming.

This may have led to a passive global warming in the mid-Paleocene as carbon from organic matter in deep-sea sediments was exhumed along the uplifted margins of continents. The carbon came from the accumulated remains of huge amounts of dead organisms; plankton, fish, and microbes. This event eliminated the "carbon sink" effects of a part of the global sea volume and enhanced the greenhouse effect as more carbon, in the form of carbon dioxide gas, found its way into the atmosphere. According to current geological theory, global warming during the Paleocene–Eocene transition was a consequence of a combination of factors, rather than a single event. These factors also include the profuse volcanism in the North Atlantic as Greenland was detached from North America and Europe; ocean warming in high latitudes, weakening the atmospheric circulation; and the increase in productivity of organic matter in the oceans.

Throughout the Eocene and Paleocene, climates across the globe were generally warm and equable, so that many places— among them northwestern North America, southern Germany, and the London

See Also

THE NEOGENE: *Separation of Australia from Antarctica*
THE PLEISTOCENE: *The Ice Age*
VOL. 1, THE ARCHEAN: *Tectonic plates; seafloor spreading and ocean ridges*
VOL. 2, THE CRETACEOUS: *The Western Cordillera of North America*

THE AGE OF MAMMALS

The Paleogene is limited to just the last one and a half percent of the earth's existence and lasted for about 41 million years. During that time the mammalian fauna of the modern age began to take shape; while many of its component animals were the precursors of modern animals that are familiar today, many were strange and extraordinary. During this time, the mammals came to fill all the ecological niches freed by the extinction of the dinosaurs. Climate and vegetation changes also gave opportunities for diversification to new groups of creatures.

PALEOGENE

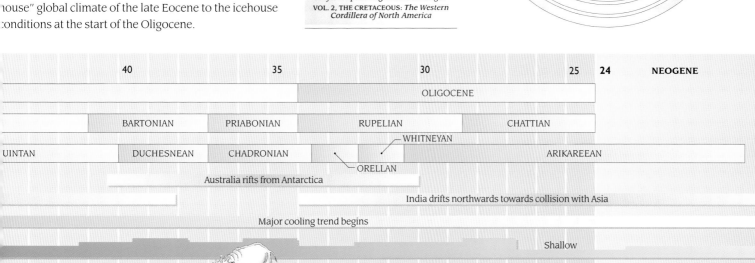

PALEOGENE

GIANT'S CAUSEWAY

(Left) The hexagonal basalt columns of the Giant's Causeway in Northern Ireland are evidence of the vast quantities of lava that poured from the fissures of the opening North Atlantic.

Basin, England—resembled the rainforests of modern Central America and Southeast Asia. Characteristic plant fossils include magnolias, citrus, laurels, avocados, sassafras, camphor, cashews, pistachios, mangoes, and tropical vines. Temperatures remained stable at about 68–77°F (20–25°C), varying by only 7–15°F (5–10°C) per year. Deltas, forests with tall canopies, and trees decorated by vines and lianas all existed in these habitats. Broad floodplains bordering meandering rivers were in evidence and would have been very much like areas such as the Amazon delta and basin of today. Fossils of the still living Nypa palms and cycads (sego palms) demonstrate the lush conditions.

Grasses had been present by the Late Paleozoic, but it was only in the late Oligocene that they began to flourish beyond their original niche in swamps and woodlands. This depended on the ability to survive intensive grazing by animals, which in turn required a quick and dependable method of reproduction. When grasses evolved that could be pollinated by wind instead of insects, they spread rapidly to form the great prairies of the world.

DURING the Paleogene the ancient Tethys Sea began to close as the African plate pushed up against Europe. This east-to-west body of water extended from southern Europe to the area of China, and was the precursor of the (much smaller) modern Mediterranean Sea. The Tethys Sea was a major oceanic realm and its effects on global climates must have been significant. The plate tectonic forces driving the closure of the Tethys eventually gave rise (in the subsequent Neogene period) to the great chain of Eurasian mountain ranges to which the Swiss Alps belong. Also part of the chain are the Pyrenees of Spain and France; the Pennines of Italy; the Carpathians of

Africa pushed up against the edge of Europe, closing the Tethys Sea, building mountains from Spain to Tibet, and forming the Mediterranean.

southeastern Europe; the Caucasus of Eurasia; the Himalayas of south Asia; and the Atlas of north Africa.

Rifting, in contrast, was not a very widespread event in the northern hemisphere during the early Paleogene. About 65 to 60 million years ago, the individual continents of Laurentia and Baltica, which had formerly made up the giant northern supercontinent Pangea, were spreading apart at only some specific areas within 30° latitude of the North Pole. This site, between what is now Greenland and Europe, was the opening of the great Mid-Atlantic Ridge, a geological feature of the Atlantic Ocean that is still prominent today. The Atlantic was narrower than it is in modern times, but seafloor spreading as the mid-ocean rifts expanded caused the ocean to widen.

In the north Atlantic, as a result of the rifting, huge volcanoes were formed far out into the ocean, including one that became the modern island of Iceland. The regions that are now Northern Ireland and Scotland experienced intense volcanic activity, with lava flows 1mi (1.5km) thick forming extensive basalt plateaus. One of the flows that occurred in this region formed the Giant's Causeway in Northern Ireland. The strange, regular polygonal jointing that gives the characteristic shapes to the columns in this famous rock formation was due to the cooling of the lava flows at a very uniform temperature. Evidence for the same activity in eastern Greenland is found in volcanoclastic deposits (rocks composed of the cemented, fragmented remains of volcanic rocks) sandwiched in between nearshore marine sediments that are widespread in northwestern Europe.

During the Cretaceous, the Arctic Ocean basin had been largely isolated from the Atlantic because North America, Greenland, and Europe still formed one huge landmass. This isolation ended as continental rifting in the Atlantic continued to spread north. During the Paleogene the Mid-Atlantic Ridge gradually split along two forks that separated Greenland from North America on the west and from Europe to the east. At this point Greenland became a large island completely detached from the remainder of Laurentia.

The eastern border of what is now North America was feeling the powerful tensional effects of its separation from Europe and Africa as the Atlantic widened from the

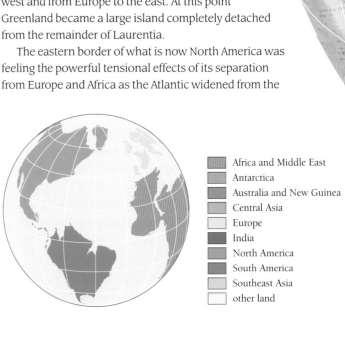

Africa and Middle East
Antarctica
Australia and New Guinea
Central Asia
Europe
India
North America
South America
Southeast Asia
other land

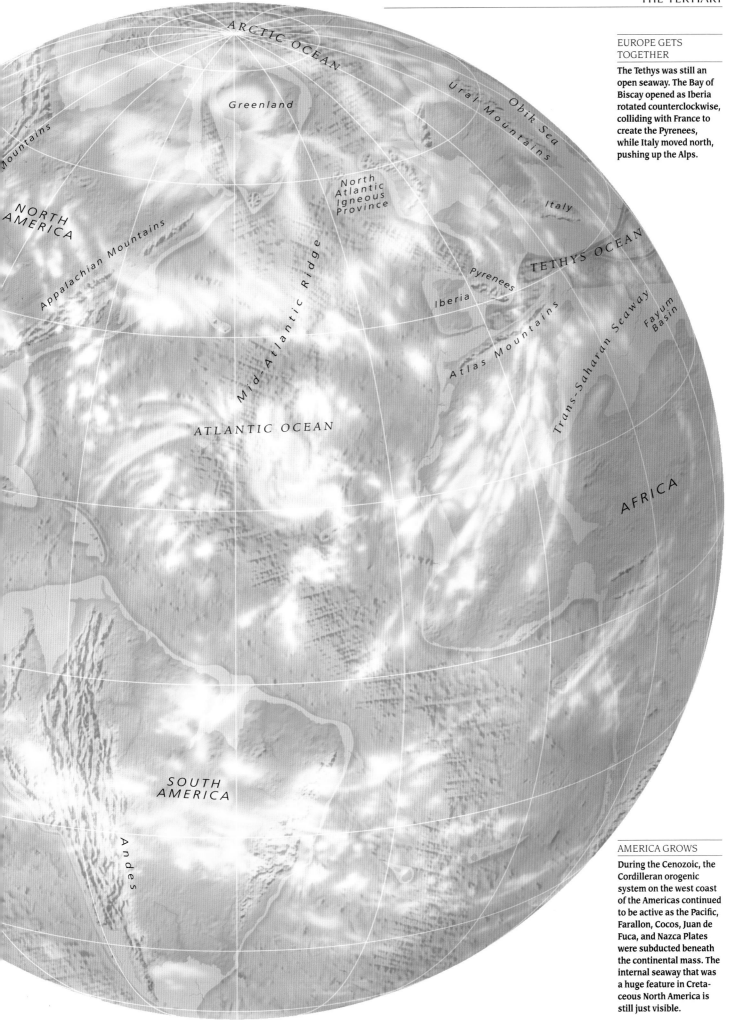

EUROPE GETS TOGETHER

The Tethys was still an open seaway. The Bay of Biscay opened as Iberia rotated counterclockwise, colliding with France to create the Pyrenees, while Italy moved north, pushing up the Alps.

AMERICA GROWS

During the Cenozoic, the Cordilleran orogenic system on the west coast of the Americas continued to be active as the Pacific, Farallon, Cocos, Juan de Fuca, and Nazca Plates were subducted beneath the continental mass. The internal seaway that was a huge feature in Cretaceous North America is still just visible.

mid-ocean ridge. In the west, tectonic compression as the North American plate overrode the Pacific plate gave rise to the Laramide orogeny, which continued well into the Cenozoic era. This event produced many of the geological features seen in the Rocky Mountains in the Cordillera region of North America. This subduction zone extended all the way along the length of South America, where the Andes continued to form.

THE LATE Eocene saw the final breakup of the continent of Gondwana with the separation of Australasia from Antarctica, and their subsequent isolation, as both moved in opposite directions. Geographical transformations elsewhere on this side of the world were brought about by the development of new subduction zones in southeast Asia, Japan, and the South Pacific, extending all the way to the Pacific coast of North America. Island arcs formed in the area from volcanic activity that gave the area its nickname, the Pacific "Ring of Fire," an epithet still valid today. The South China and Philippine seas appeared in the region at this time. In the late Paleocene, about 57 million years ago, there was still a significant seaway between the drifting Indian island and the southern margin of Asia, though mountains were already rising between the two. This seaway was part of the ancient Tethys sea. Farther east, the Turgai Strait linked the Tethys to the Arctic along the Ural Mountains.

As Australia rifted away, Antarctica became isolated over the South Pole and surrounded by a cold circumpolar current.

As the Australian continental plate drifted north, it lost contact with the landmass of Antarctica, allowing the formation of a circumpolar Antarctic current. In essence this event was the beginning of our modern climatic environment. Even before its separation from Australia, Antarctica had been centered over the South Pole but had remained warm because its shores had been lapped by warm waters from lower, subtropical latitudes. Plate tectonic movements caused an alteration of wind and oceanic currents, which stopped warm equatorial waters from getting to Antarctica because the cold ocean waters around the great southern continent began to flow in a circular direction around it, blocking the warmer waters from the north. This new distribution of cold waters was fully established in surface waters by the Oligocene.

In response to this cooling, the first sea ice began to form near the end of the Eocene and waters of near-freezing temperatures sank to the depths (cold water is denser than warmer water and so sinks). The freezing waters that sank to the deep sea around Antarctica then spread Northwards forming the psychrosphere—the deepest zone of the ocean, characterized by near-freezing water. The psychrosphere has had a major impact on deep-sea life and even now is a major physiological barrier to most oceanic life. At this time a major extinction is seen in the fossil record of deep-dwelling foraminiferans, microscopic calcite-shelled protistan

animals that are abundant at various depths in the oceans. Other deep-dwelling marine organisms such as mollusks show extinctions as well.

These events were spread over a long period of time that extended from the mid-Eocene through to the mid-Oligocene. A sharp decline in animal life coincided with a major sea regression as the volume of ice over Antarctica increased rapidly.

ANOTHER change in marine ecology as a result of the formation of the psychrosphere was that of the re-expansion of scleractinian corals (which include all modern corals, as opposed to the rugose and tabulate coral groups that went extinct at the end of the Permian period). These stony or "true" corals survived the cooling episode to become the significant reef-builders of modern times. Larger marine animals were also affected, though not necessarily for the worse. The first whales appeared during the early Middle Eocene in the subtropical waters of the eastern Tethys. They were aquatic, carnivorous mammals and the precursors of the modern toothed whales. Towards the end of the Oligocene, as the world became colder, the baleen whales

The cooling that took place during the transition from the Eocene to Oligocene directed much of subsequent evolution.

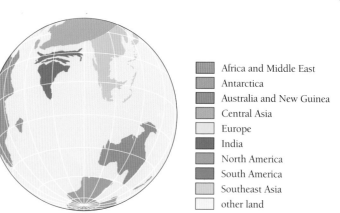

Africa and Middle East
Antarctica
Australia and New Guinea
Central Asia
Europe
India
North America
South America
Southeast Asia
other land

ASIA

Ancestral Himalayas

INDIA

INDIAN
OCEAN

Southeast Indian Rise

Java Trench

New
Guinea

Kerguelen
Landmass

AUSTRALIA

Antarctic–Pacific Rise

ANTARCTICA

PALEOGENE

SUBDUCTION IN THE PACIFIC

New subduction zones appearing at the margins of the Pacific basin gave rise to an almost continuous volcanic arc system, forming what is known as the "Ring of Fire" around its rim.

AUSTRALIA ISOLATED

The Australian continental plate rifted away from the Antarctic Plate during the first half of the Paleogene and has drifted 500mi (800km) to the north since that time. Australia's movement created a series of deep ocean trenches and island chains which are now the Indonesian archipelago.

evolved, with their system of horny plates used to filter the planktonic organisms that would have thrived in the colder waters. Like all mammals, cetaceans are warm-blooded; to maintain their temperature they lay down fat, in the form of blubber, below the skin. Larger species—such as the baleen whales—have an advantage with their more favorable surface-to-volume ratio.

SOME 2mi (3km) below sea level, the Mid-Atlantic Ridge marks the boundary between the American and African plates. It extends nearly 10,000mi (16,000 km) in length from the Arctic Circle to the southern tip of Africa, and 1,000km (1600km) in width from its position at an equal distance from the continents east and west of it. At the crest of the ridge, a narrower zone about 50 to 75mi (80 to 120km) is the site of the underwater volcanic activity that causes the seafloor to "spread." Magma welling up from the crust is pushed away from the sites of eruption, building up new seafloor. In this way the basin of the Atlantic is estimated to be widening by about 0.5 to 4in (1 to 10cm) every year.

The Mid-Atlantic Ridge is a ten-thousand-mile-long underwater mountain range, created as volcanic material rises up from the mantle.

GRABEN STRUCTURE

Faults bounding the sides of a down-dropped graben extend below to the underlying magma chamber and form connections to the surface through which lava is extruded. These lava flows successively form new floors to the graben as the lateral extension continues, forming new oceanic crust.

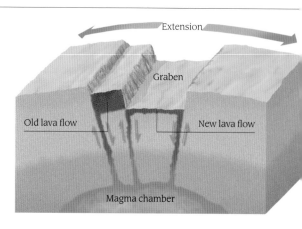

Iceland is a volcanic island of the Mid-Atlantic Ridge that has built up until it has risen above sea level, a boon for geologists. The "island," formed entirely from oceanic crust, is a contiguous part of the undersea mountain chain, rising more than 1.5mi (2.5km) higher than the rest of the ridge—enough to raise it above sea level and dry out the basaltic rock. Other volcanic islands along the ridge that have also risen above sea level are Ascension Island, St. Helena, and Tristan da Cunha, though none of these sits exactly on the Mid-Atlantic Ridge as Iceland does. All are associated with "hot spots" where molten material rises from the mantle, as much as 375mi (600km) below the surface. The hot spot directly under Iceland is responsible for the great buildup of lava that

MID-OCEAN RIDGE

The mid-ocean ridge of the North Atlantic is a vas[t] feature that runs along the seabed for several thousand miles; there is one place, however, where it is exposed on the earth's surface—on Iceland. The steep-sided valley of the Thingvellir graben (below), where the basalt formation of the Mid-Atlantic Ridge surfaces, is a striking demonstration of tectonic plate divergence.

NEW OCEAN ARMS

The huge, rugged mountain range of the Mid-Atlantic Ridge that lies in the middle of the Atlantic Ocean is the site of the formation of new oceanic lithosphere. New crust spreads laterally, at a rate of about an inch (2.5cm) a year, carrying the continents with it, as magma—in the form of extruded lavas—moves outwards from the fissure. Between 63 and 52 million years ago there were enormous outpourings of volcanic material as a mantle hot-spot stretched the earth's crust. (These masses of basalt lava are preserved today in Northern Ireland and the Inner Hebrides of Scotland.) Ruptures opened up the narrow seaways between Europe, Greenland, and North America (top map), eventually isolating Greenland as a large island and connecting the Arctic Ocean with the Atlantic. In mid-Paleogene times, the western rift became inactive, but the eastern arm continues to widen today (lower map).

ZONES OF SEAFLOOR SPREADING

The mantle plume that initiated the rifting of Greenland from Europe still exists under Iceland. The progression of the North Atlantic constructive plate margin can be traced across the island by the zones of volcanic activity and the rifting structures that are found there. The whole island is volcanic in origin; the island of Surtsey surfaced in 1963.

- zone of rifting
- volcanic deposits
- continent
- deep oceanic basin

Iceland detail
- Tertiary basalts
- Pliocene–Pleistocene basalts
- Upper Pleistocene basalts
- recent sedimentary deposits
- fissure
- ▲ volcano

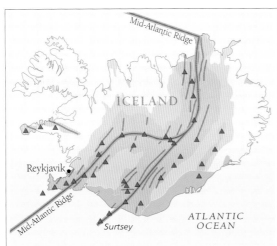

constitutes the island. It was the same hot spot that produced huge lava flows covering hundreds of miles in eastern Greenland about 60 million years ago when Greenland began to separate from Europe.

The American geologist Harry Hess, author of the theory of seafloor spreading, noted that mid-ocean ridges are characterized by deep furrows, which in Iceland are visible as features of the landscape. Valleys, called grabens, are created where there are two fault zones on either side of a central block that slips down between them; this forms the furrow. Grabens form where crust has split and new lava has risen up. Submersible craft exploring the Mid-Atlantic Ridge have discovered the floor of the narrow rift zone to be covered with pillow lavas, the globular shapes that fresh lava often takes when rapidly cooled under water. Moreover, the valley floor is covered with fissures up to 30ft (10m) wide, parallel to the ridge axis, as if the bed of the sea is being pulled apart by enormous stresses.

ALSO in the early Paleogene, but much further to the southeast, another phenomenon related to the rifting in the northeastern region of the Atlantic ocean was that the seafloor of the ancient North Sea basin began to sag, accumulating thousands of feet of sediments in the center of this geological basin. The North Sea, touching the shores of nearby continents (what is now northern Europe), deposited extensive fine-grained marine sediments in southeastern Britain and the European mainland between northern France and Denmark. These deposits show that the conditions during the time of their deposition were those of subtropical environments. Later in the Paleogene, when sea levels rose, the North Sea flooded northern Europe as far east as the Ural Mountains.

Ebb and flow of the ancient North Sea left extensive marine deposits in Paris and London. Typically of the time, these were tropical.

Two deposits from the area are particularly important in the history of European geology. Sir Richard Owen in Great Britain studied fossils from the clay of the London Basin in detail. This legacy from the late Early Eocene yielded an array of exquisitely preserved fossils such as crocodiles, turtles, sharks, fishes, mammals, and even tiny birds—350 species in all. The animal and plant remains show it to have been a tidally-influenced subtropical shoreline that was intermittently forested. It was similar to a modern-day mangrove swamp, and some of its species have relatives in modern Malaysia.

The fossils of the Paris Basin limestones had been studied by the great comparative anatomist Baron Georges Cuvier, head of the National Museum of Natural History in Paris. They show a community of fossil animals similar to those of the London Clay. Cuvier's discovery and description of the ancient tapir-like fossil mammal *Palaeotherium* ("ancient beast") was the first of its kind and of great importance to the history of paleontology.

part of the Asian plate, the resulting undersea orogeny, in association with the fore arc (a submarine feature for long distances), produced a curving line of seamounts that now show as the Javanese Islands of the Indonesian arc These include Java, Sumatra, Sulawesi, and New Guinea; they also include the volcano Krakatau, which last exploded in 1883. The Emperor–Hawaiian seamount chain, which had begun to form about 70 million years ago, began to curve as the spreading direction of the East Pacific Rise became more westerly at 43 million years ago, and new subduction zones and arc systems developed south of Japan and in the South Pacific.

The Mentawai Islands off the coast of Sumatra show a good exposure of Cenozoic strata. The rocks in these islands can be seen to have been strongly folded in the Early Cenozoic and then again during the Miocene. Terraces of uplifted coral reefs show that these islands have recently been rising; that they are still doing so is indicated by the geology of the Nicobar and Andaman Islands that extend towards Burma, where these islands become a continuous land feature and pass into the high mountain range of the Arkan Yoma west of the Irrawaddy delta. Far to the east of Java, out in the Pacific Ocean, the ridge of the Indonesian island arc rises as the island of Timor. This island is particularly interesting for the great height to which its successive coral reefs have been upheaved, some of which now stand well over 3900ft (1200m) above sea level. These reefs record the

Ocean crust is created along mid-ocean ridges at rift zones and "destroyed" in deep-sea trenches where plates descend into the asthenosphere, to be re-melted and re-cycled through the mantle. Such areas are called sub-duction zones, and the majority of them are found around the margins of the Pacific Ocean, where they formed as a result of the oceanic lithosphere sub-ducting beneath the conti-nental plates of Asia and the western edge of the Americas. (The Philippine plate, one of the smaller plates in the world, is entirely surrounded by subduction zones, crushing it between the Pacific and the eastern margin of Asia.) The entire Pacific basin is characterized by intense volcanic and earthquake activity, earning it the nickname "the Ring of Fire." The same belt continues along the underside of Southeast Asia, at the northern edge of the Australian plate, and extends up into the Asian mainland, following the line of the Himalayas and across to the southern Mediterranean.

Most of the world's major subduction zones are found in the Pacific basin, forming a great "Ring of Fire" around its edges.

All of this activity began about 60 million years ago during the Paleogene as Australia began to pull away from Antarctica, taking with it New Zealand and New Guinea. India, a small continental fragment, was already heading north on its rendezvous with Tibet. As the embayed central portion of the northern rim of the Aus-tralian plate impacted with the projecting southwestern

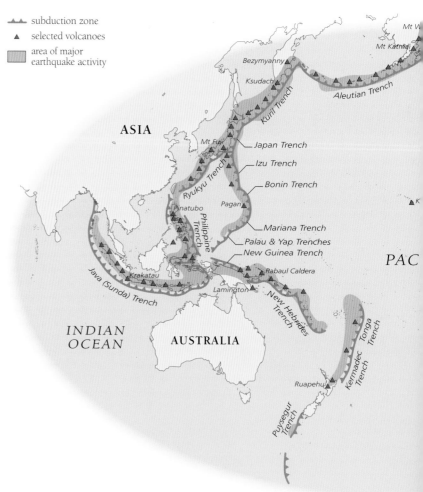

▲▲ subduction zone

▲ selected volcanoes

area of major earthquake activity

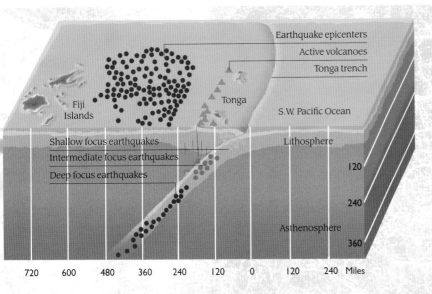

BENIOFF EARTHQUAKE ZONE

When the oceanic lithosphere sinks into the earth's interior at subduction zones, strains are set up which generate earthquakes. In 1954 the seismologist Hugo Benioff discovered that the differing depths of earthquakes (shallow, intermediate, and deep) were associated with the angle or "dip" of subduction zones, lying on a sloping plane corresponding to the plane of the descending plate. This characteristic earthquake pattern is very well illustrated by the Tonga region of the South Pacific. Away from the Tonga Trench, in a roughly northwest direction, the earthquake foci become increasingly deeper as the lithosphere descends. Earthquake zones associated with ocean trenches are known as Benioff zones and as a general rule they show that the angle of a subduction zone increases with increasing depth, while the magnitude or strength of earthquakes decreases with increasing depth. Consequently, it is shallow earthquakes that are the most powerful and dangerous; in ocean–continent destructive margins these tend therefore to be located nearer the shores of continents.

HOT-SPOT VOLCANO

Kilauea volcano (left) is one of the five shield volcanoes that form Hawaii, part of an island chain lying in the middle of the Pacific Ocean, far from any plate boundaries. It is a hot-spot volcano, producing the runny basaltic lava found at constructive margins.

changes in height of the islands during their development and also sea level changes.

The unity of this very long chain of islands, mountains, and submarine ridges is also demonstrated by its coincidence with a belt of negative gravity readings from the submarine trench that are particularly strong underwater. This anomaly was discovered in the 1920s by the pioneering Dutch geologist F.A. Vening Meinesz. By submerging his gravity-reading equipment, he overcame the distorting "pendulum" effects of the waves and eventually charted a well-defined negative gravity belt along almost 2500mi (4000km) of the Indonesian Island arc. He inferred this to represent a down-buckling and consequent thickening of the oceanic crust to form a mountain root.

Modern geophysical analysis, however, shows that the

oceanic crust beneath these gravity anomalies is of the same thickness as transitional regions between the continental and oceanic margins.

Back-arc spreading—the extension and spreading of the ocean floor situated behind an island arc—may occur at convergent plate margins, and there is evidence to suggest that this is the result of very complex convection eddies taking place in the asthenosphere located above a subducting plate, or by the pulling away of the adjacent plate. Regional tension and normal faulting result from the effects of the rising mantle material, and also from stretching of the crust as it is "rolled over" by the friction and grip of the crust of the subducting plate. This back-arc spreading takes place behind the zone of subduction and, as a result of this tension, sags. It is now situated at a lower level than the surrounding continents and forms the incipient stages in a sea. Back-arc basins may attain huge dimensions. The floors of back-arc basins are generally young and sediments are thin; exposed rocks include fresh basalts and they cease to become active regions of spreading after a period of time—perhaps about 10 or 20 million years. Heat flow is often high but there is no well-defined ridge, as in the case of the Atlantic Ocean and its prominent mid-ocean ridge. Back-arc spreading, and the resulting basins, are considered to have been common throughout much of the Earth's history. The Sea of Japan is a back-arc marginal basin formed by spreading between Japan and the Eurasian margin and is now inactive; it appears to have begun forming during the early part of the Cenozoic.

The formation of the Arabian and Indian Oceans are very closely tied together. The Indian Ocean formed when the Indian plate eventually contacted the southern edge of the Eurasian plate; and the geographical region now occupied by the Arabian Sea—between the western edge of India and the eastern edge of the Arabian peninsula—came into existence. Ophiolites in the Gulf of Aden are dated to the Early Cretaceous and show that this is the time when the formation of the gulf began. This sequence was complete by about 20 million years ago.

PACIFIC RING OF FIRE

Where the edges of oceanic plates subduct beneath the margins of continental plates, deep trenches form as the oceanic lithosphere is carried down into the mantle. At depth it melts and water locked up in it also melts the overlying mantle. The molten rock collects, and being less dense, rises through the continental crust, erupting explosively at the surface. The huge plates that comprise the Pacific Ocean basin are bounded by subduction zones—and hence volcanoes and earthquake activity. Because of this, the resulting lines of volcanic activity are in the form of a gigantic arc that extends around most of the Pacific basin—the "Ring of Fire."

IN EUROPE, from the middle of Paleogene times, there were major changes to the physical geography. This was the origin and formation of the Alps mountain range.

Europe's physical geography altered dramatically as Africa pushed up against the European plate, raising the Alps.

Geologically speaking, the Alps are rather young mountains. They were formed as a result of the northward movement of northern fragments of the African plate, causing impact with the European plate. The ancient Alps were part of a long string of mountain chains that ran from west to east along the southern border of Eurasia, from the Pyrenees on the border of France and Spain, the Carpathians in eastern Europe, and the Himalayas in Asia. The Atlas Mountains are their counterpart in northern Africa.

The highest of these ranges are the Himalayas, but all of these mountains formed as a consequence of the northward movements of Gondwana. They are positioned along what was until Paleogene times a part of the southern margin of the Eurasian landmass, bounded by the ancient Tethys Seaway. This history has been confirmed by the discovery within the Alps of ophiolites

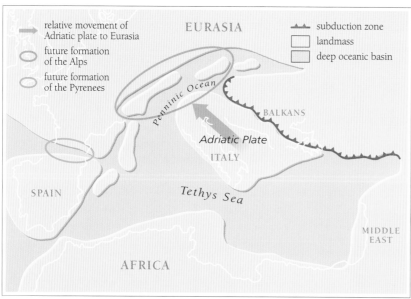

ORIGIN OF THE ALPS

As the Adriatic plate moved north it sutured Italy to Eurasia, closing the Penninic Ocean and uplifting the Alps. A similar collision by the Iberian peninsula created the Pyrenees. The Alps form a typical mountain chain with each zone representing a gradient of deformation. The igneous core lies in the Southern Alps, the Pennine zone is the metamorphic belt, while further north is the fold-and-thrust belt.

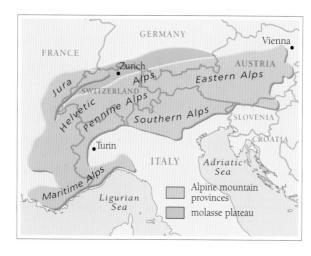

that represent Mesozoic and Early Cenozoic oceanic crust. These Alpine ophiolites occur mainly in the Pennine Alps between Italy and Switzerland—for example, at the base of the famed Matterhorn.

Interestingly, in most of the zone of orogenic activity at the junction of Africa and Europe, the African continent has not been sutured to the Eurasian continent. As a result, there is considerable seismic activity in this region as the mountain-building continues. This activity is apparent as earthquakes, especially in southern Italy and Turkey. Zones of seismic activity plotted on a map follow the line of mountain-building across the Mediterranean and the Middle East. Faults in these regions closely follow the earthquake zones.

THE ACTIVE tectonic zone running through the Mediterranean region is a reflection of the structure of the earth's crust in this area. Here, many small plates of lithosphere (the outer, rigid shell of the earth, including the crust) have been caught between the African and Eurasian shields, and these microplates have been pushed against one another and sandwiched between the African and Eurasian plates to produce a mosaic of tectonic activity. The Alps had their geological origins among the Iberian, Corsican, Sardinian, and especially the Adriatic microplates, which attached themselves to southern Europe. The Adriatic plate had originally been attached to the Eurasian continent in the region of the

Microplates carrying the Iberian and Adriatic peninsulas, as well as Corsica and Sardinia, contributed to the Alps' rise.

modern Balkans, and it carried the entire peninsula that eventually became known as Italy. About 45 million years ago, as the northern edge of the Adriatic plate was crumpled against the southern margin of Europe, the edge of the plate rode over the margin of the Eurasian craton. As the crust was thickened and buckled by this action it also received an accumulation of igneous rocks above the subducted plate. This complex crustal structure is the basis for the formation of the Alps.

During the first stages of the Alpine mountain formation, the emerging folds of crust were separated by patches of sea. (The Penninic Ocean had appeared in the region north of the Tethys, caused by rifting during the Triassic period.) Highly distinctive sediments accumulated in these folds; these were dark siliceous shales (indicating deep waters and the deposition of sand) and sandstones in random arrangements that were deposited between elongated deepwater submarine banks. These characteristic uplifted sedimentary rocks in the Alps are known to geologists as flysch. By Oligocene times, compression from the south had caused enormous recumbent (turned sideways) folds to rise as mountain arcs out of the old seaway and to slide over the land to the north. North of these complex folds lay a geographical depression that received these land-derived clastics (aggregated rocks), called molasse. A great plateau of this material lies on the northern edge of the Alpine system.

The Alps are formed into several mountain ridges that run more or less parallel to each other in an east-west direction. From the north these ridges are known as the Jura Mountains, the Helvetic Alps, the Pennine Alps, and the Southern Alps. The Helvetic Alps of Switzerland contain a particularly large volume of flysch rocks. The

oldest flysch deposits now found in the Alps were deposited within the Penninic Ocean, which, during the Mesozoic, temporarily existed between the southern margin of the Eurasian plate and the north edge of the advancing Adriatic microplate. The collision between the two plates destroyed this ocean during the Eocene epoch, but it left deep marine waters standing to the south of the Alps, where younger flysch deposits accumulated. The Alpine orogeny continued to the Late Miocene, between 10 and 5 million years ago.

MOUNTAIN-BUILDING on the western margin of North America, which had begun at the end of the Cretaceous, continued into the Eocene. The ranges created by the Laramide orogeny extended from Mexico up to Canada. Tectonic activity also spread eastward, as far as the Black Hills of South Dakota. Between the local mountain ranges, such as the Wasatch and the Front Range of the rockies, lay broad basins. These received not only the erosional detritus from the nearby mountains, but also the discharge from their streams. Sediments and water accumulated in the subsiding areas throughout the Paleocene–Eocene, forming lakes and swamps. These basins were the Bighorn, Uinta, Washakie, and Green River basins. The Uinta Basin is the deepest of the Eocene lakes in this region. The Green River formation is notable for its rocks, which consist of almost 2000ft (600m) of freshwater limestones and very

In western North America, mountains had been raised from Mexico to Canada. Great lake basins lay between the ranges.

NEW PEAKS

The relative geological youthfulness of the European Alps (above) is reflected in their great height: this is because they have had less time to erode away than older mountain chains.

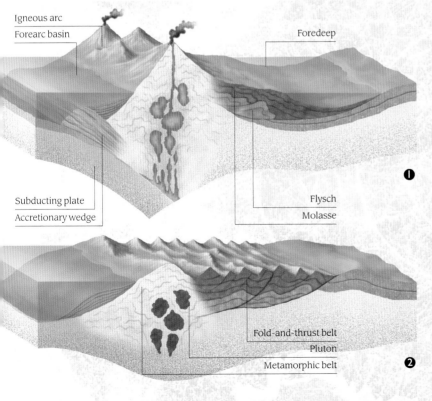

Igneous arc
Forearc basin
Subducting plate
Accretionary wedge

Foredeep
Flysch
Molasse

❶

Fold-and-thrust belt
Pluton
Metamorphic belt

❷

THE ANATOMY OF MOUNTAIN CHAINS

Mountains form at the collision zones of tectonic plates, either between two continental plates (as in the Alps and Himalayas) or between a continental and an oceanic plate (as in the Andes). As a subducting oceanic plate slides beneath a continental craton, immense shearing stresses are set up and slices of the surface are scraped off to accumulate in the angle between the two. This debris forms an accretionary wedge in front of the newly-forming mountain range (1). Magma from melting subducted material rises through the continental crust, forming a chain of volcanoes—the igneous arc—which is the core of the incipient mountain range; some magma crystallizes into large bodies of igneous rocks called plutons. Sediments accumulate in the sea on top of the accretionary wedge, which is known as the forearc basin. These sediments are mostly derived from the land. Beyond the mountains the crust is downwarped to form a foredeep. At first the foredeep fills with marine flysch deposits but as sediment transport from the mountains increases the sea retreats and flysch deposition is replaced by non-marine sediments called molasse.

The subducting plate is pushed into the region of metamorphism where extreme heat and pressure cause chemical and structural changes to the rocks. These metamorphosed rocks lie at the very deepest roots of mountains and, when eventually exposed in a mature mountain range, represent the most ancient rocks that are visible to geologists (2). This region forms a well defined area and is termed the metamorphic belt. Inland, the crumpled-up rocks are folded and ride over each other in great sheets, in an area known as the fold-and-thrust belt.

academic) John Bell Hatcher discovered 23,000lbs (10,400kg) of immense animal bones. These bones were eventually recognized and named as those of enormous elephantine mammals called titanotheres ("titan beasts"). Graveyards have been found in ancient river channels, showing characteristic evidence of flash floods.

Titanotheres—properly known as brontotheres ("thunder beasts")—were browsers that ranged in size from that of a large pig to a modern elephant. They have, however, no real modern-day analogues, for they were equipped with ossicones (as in living giraffes) that were located on the top of the snout, above and behind the eyes, and on the cheeks. Some forms such as *Bathyopsis* sported huge, downward-curving canine tusks similar to some pigs today. The combination of skull horns, huge size, and lethal tusks probably made the adults of these animals almost invulnerable to the predators of the Eocene, such as the creodonts, which were much larger than the modern grizzly bear.

Titanotheres, creodonts, early "true" carnivores, early members of the horse family, ancient rhinos, and even mysterious groups such as the taeniodonts and tillodonts were in evidence, as were early primates, insectivores, and the archaic carnivores known as condylarths.

fine laminated shales. The laminations in these basins are varves—thin sedimentary layers or pairs of layers that represent the deposition of a single year. From these it has been calculated that the Green River sediments were deposited over more than 6.5 million years.

COLORFUL CLIFFS

Rocks of the Wasatch Formation in Wyoming were deposited by Eocene rivers. They preserve many mammal fossils.

THE WHITE River Formation is particularly well known for its fossils of aquatic marginal subtropical fauna. Flash floods in the Eocene deposited floodplain sediments that contain numerous fossil mammals, drowned and then preserved in sediment as it dried out. Large mammals that lived on the plains were also preserved, presumably due to being washed in during flash floods and also from dying near to the water's edge.

Some 18 million years after the extinction of the dinosaurs, mammals were thriving in and around the lakes of the Wyoming region.

Much of what is now known about the evolution of mammals in the Cenozoic of North America came from these rocks.

Many other types of fossils were preserved in the lakes. Plant fossils confirm that the area was a subtropical environment, as do the presence of numerous fossils of frogs, pond turtles, lizards, boa constrictors, and crocodiles. These animals lived along the shores of the lakes and show a thriving and diverse lakeside vertebrate community, although the wonderfully preserved remains of freshwater fish are of particular interest.

The sediments in the Paleogene lakes in and around Wyoming were first studied during the nineteenth-century expeditions of O.C. Marsh of Yale University. In 1866 the field collector (and subsequently renowned

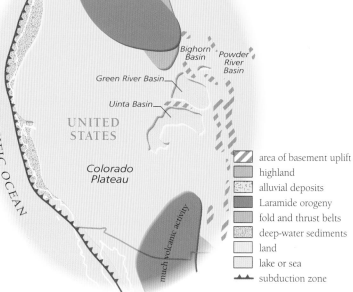

area of basement uplift
highland
alluvial deposits
Laramide orogeny
fold and thrust belts
deep-water sediments
land
lake or sea
subduction zone

THE WYOMING LAKES

At the end of the early Eocene the angle of the west coast subduction zone changed from steep to shallow dipping, producing a band of volcanism and fold-and-thrust belts 900mi (1500km) inland. Farther east, sections of Precambrian crystalline basement rock were uplifted in north-south lying ranges. As water streamed off these elevated areas, basins in front of them formed lakes, which soon filled with outwashed sediments from the rapidly eroding mountains.

EOCENE HERBIVORES

The lakes of Wyoming and Utah are famed for their fauna of extinct ungulates, beautifully preserved in the fine sediments. During the Eocene the first hoofed mammals evolved to fill a wide variety of niches. The largest mammals at that time were the dinocerates ("terrifying horns"), which included the rhino-like *Uintatherium*. *Eohippus*, the "dawn horse," also appeared at that time.

MAMMALS had lived through the Mesozoic era as tiny burrowers and tree-dwellers. They existed in the shadow of the dinosaurs for 140 million years, until the end-Cretaceous extinction provided an enormous evolutionary opportunity. The evolutionary driving force that was behind a riot of adaptive evolution of the early Paleogene produced many land mammals that are no longer around and that look immensely different from those of today; most have no modern anatomical or ecological analogues whatsoever, which poses a real obstacle to investigating their individual lifestyles.

Mammals of the Paleogene were strange-looking to modern eyes and have no analogous forms among modern species of mammals.

Mammals during the Paleogene were bizarre by modern standards. Among the characteristic groups of primitive mammals from this period were the condylarths, which included abundant forms such as the huge tusked pantodont *Coryphodon*; small rabbit-deer-like hyopsodonts; and phenacodonts. They were stockily-built animals without claws, but a blunt hoof on each toe. Condylarths evolved some predatory types such as the superficially wolf-like *Mesonyx* and the immense "hyena-bear" *Pachyaena*. With hoofed toes and teeth that could rip, these impressive predators may be thought of as "wolf-sheep." They belonged to an assemblage of animals known as mesonychids, characterized by their five-toed feet—with hooves, rather than claws—and specialized front teeth.

But they did not have the niches of Paleogene predators all to themselves. Creodonts, the ancient cousins of true carnivores, were abundant at this time. They were more specialized than the giant mesonychids, having teeth that could cut, rather than rip flesh. Creodonts were among the more bizarre combinations of different animals, some resembling gigantic hybrids of bear, hyena, and lion. Creodonts such as *Patriofelis* and *Oxyaena* were low-slung, stocky animals with heavy tails, powerful jaws, and mouths full of slicing teeth. These huge early carnivores fed on the various ungulate-like (hoofed) browsers that were abundant in the subtropical Paleogene forests.

These strange herbivores included perhaps the oddest Paleogene mammals of all: the extraordinary taeniodonts, which were essentially part bear and part rodent. Taeniodonts such as *Psittacotherium* and *Stylinodon* possessed body proportions rather like that of an unusually thick-set bear, and had huge digging claws on the front feet, but the head was exceedingly deep and had an immense pair of open-rooted, extra-robust gnawing teeth, in which the bands of enamel gave rise to their name: taeniodont means "banded tooth."

SKULL BUMPS

The extraordinary 24in (60 cm) skull of the huge brontothere *Uintatherium* shows its array of horns, perhaps used in mating rituals, and its huge canine tusks, which were possibly used for fighting.

1 *Dolichorhinus* (a brontothere)
2 *Hyracotherium* (a very early horse)
3 *Hyrachyus* (an early rhinocerotid)
4 *Meniscotherium* (a condylarth)
5 *Amynodontopsis* (a small "running" rhinoceros)
6 *Uintatherium* (a brontothere)
7 *Phenacodus* (a condylarth)

PALEOGENE

DIETARY ADAPTATION

Bears and pandas are descendants of carnivores that also were ancestors to the dog, raccoon, and weasel families. While the cat branch of the carnivore family tree became specialized meat-eaters, the dog branch moved towards opportunistic omnivory and, in some animals such as the Kinkajou, near-vegetarianism. Bears also rely heavily on fruit and vegetables but, in general, their omnivorous diet has led to great success; bears are found on most continents, though there are just eight species. (Specializations tend to be fragile since they have evolved for one particular use; generalists can make a living off most food sources). The opportunism of some bears today (right) brings them into conflict with humans.

IN EVOLUTIONARY studies it is important to attempt to discover something about the way of life of extinct organisms. In the case of taeniodonts, any attempt to establish a modern analogue is flawed: there is no bear-like gnawer and there is no gnawing bear. They appear to have originated from insectivorous ancestors close to *Cimolestes*, an ancestral carnivore from the late Cretaceous. Taeniodonts occupied their bizarre niche of giant rodents along with a group called tillodonts. Tillodonts such as *Trogosus* were less powerfully built than taeniodonts and did not have such deep skulls, but their huge front teeth and the anatomy of their skulls gave them the appearance of gigantic rabbits. The subtle anatomical differences between tillodonts and taeniodonts probably meant that they fed on somewhat different food items; consequently there was probably less overlap in their diet and feeding habits than might seem the case. Taeniodonts had truly immense incisors, powerful jaws, and clawed feet that could have been used to dig up the roots and tubers that formed their diet. Fossils of both groups have been found in the same sediments; differences in size and dentition suggests that they lived side-by-side without competing for the same food. This is known as trophic partitioning.

Giant rabbit-like creatures such as **Trogosus** *shared the niche of giant gnawing rodents with the taeniodonts.*

The rodentlike adaptation, characterized by large gnawing front teeth, figured largely in another group of highly successful mammals, the multituberculates. Multituberculates ranged in size from a mouse to a lynx and had a wide range of ecological niches that spanned everything from burrower to arboreal gnawer. The skulls of the smaller ones looked a little like that of squirrels, as did their skeletons, but one group, the taeniolabids ("banded lips"), had chunky beaver-like heads and tough incisors. The name multituberculates arises from their

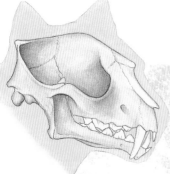

Dinictis (carnivore)

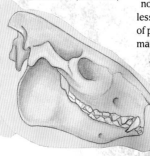

Enteledon (omnivore)

Merycoidodon (herbivore)

TEETH AND THE RISE OF MAMMALS

To paleobiologists, the one great feature that distinguishes mammals from all other vertebrates is the complex structure of their teeth and jaws, and much of the evolutionary success of mammals is due to the remarkably adaptive nature of their feeding apparatus. Mammals have complex teeth that occlude (interlock) in an extremely precise fashion, to give wear facets that are so predictable as to be practically diagnostic for some species.

The first incipient signs of the mammalian dentition was evident among the synapsid ("mammal-like") reptiles of the Triassic period. In these very advanced—for their time—mammalian precursors, wear facets had begun to appear as their teeth developed cusps and bumps on them. The teeth of synapsids had changed from a simple homodont dentition (where all the teeth are the same shape) to a heterodont dentition (where teeth are differentiated into, for example, incisors and canines) way back in Permo-Carboniferous times with animals such as *Dimetrodon*. Later, in the Triassic, the cynodont precursors of mammals added close-packed tooth occlusion to heterodonty and also developed novel arrangements of their jaw-closing musculature. Nevertheless, these animals still had not attained the mammalian condition of precise occlusion. This finally happened with the earliest true mammals in the Triassic and Jurassic, such as the shrew-sized *Morganucodon*. Jurassic multituberculate mammals developed their dental apparatus in the rodentiform (rodent-like) anatomical direction; but mammalian dental precision finally reached its modern level in the Jurassic and Cretaceous mammals that were themselves the precursors of Cenozoic large-bodied mammals.

In the 65 million years of the Cenozoic era, following the demise of the dinosaurs, a riot of mammalian dental evolution took place. After a slow start, tooth shapes became highly complex. Most early herbivores were browsing animals that fed on the leaves of trees and shrubs. Their molar teeth developed strong rounded cusps giving a lumpy crushing area on the biting surface. They are referred to as bunodont ("bun") teeth in recognition of their shape. The bunodont dentition has proved to be extremely successful ever since; it is found in a number of living mammals such as certain pigs.

Another type of molar tooth among Paleogene ungulates had sharp crescent-moon shaped cusps on the contact surfaces. This design is termed selenodont ("moon-toothed"). It was much less common among Paleogene mammal faunas than the bunodonts; however, selenodont dentition is superbly designed to grind up tough grass. When grazing animals evolved in the Oligocene in response to the expanding grasslands, the selenodont dental adaptation became very widespread.

It was during the Oligocene that another fundamental dental design was seen for the first time: the appearance of true rodents. A "rodentiform" dentition had been in evidence from the Triassic period onward, since the tritylodont cynodonts and multituberculates. However, neither of these animals had either the complex grinding molars of true rodents, nor the self-sharpening edges of the elongate incisor teeth that are so characteristic of rodents today. The adaptation of self-sharpening incisors is in part responsible for the incredible success of the rodents since the Oligocene epoch.

Dental innovation among the carnivores is represented by their characteristic molars that have become specialized for the efficient slicing and cutting up of flesh. These are the blade-like carnassial teeth; the most rearward upper premolar shearing against the most forward lower molar in a scissor action. This development accounts for the success of true carnivores as top predators. The canines in true carnivores always remained large and dangerous, although there were huge differences between the various groups.

TREETOP BROWSER

Indricotherium provides an example of how ecological niches (defined by the animals that occupy them) are filled after the disappearance of the occupier. Several million years before the appearance of this animal, the niche of "long-necked superheavyweight browser" had been filled by the sauropod dinosaurs. After their extinction, a few mammals came close to achieving this adaptation, but only *Indricotherium* succeeded completely. It stood 26ft (8m) high but belonged to the rhinoceros family, rather than elephants or giraffes.

molar teeth, which were composed of columns of enamel giving a rough surface for grinding food. The multituberculates were the group with the longest evolutionary record of any mammalian group, arising 160 million years ago in the Lat Jurassic and only going extinct at the end of the Paleogene 40 million years ago. This represents a span of 120 million years, only 30 million years less than the success achieved by the dinosaurs.

The disappearance of the dinosaurs at the end of the Cretaceous had opened up many habitats and ecological niches, at the end of one of the most single-animal dominated ecosystems ever. There followed a rapid burst of mammalian adaptive evolution—much of it in the structures of the jaws and teeth—as these previously insignificant creatures evolved to take over the vast range of now unoccupied niches. Many of the anatomical 'results' may be informally viewed as 'evolutionary experiments' and this accounts for the sometimes bizarre forms that appeared. Many of these adaptations bear no resemblances to modern mammals and are quite mysterious and extraordinary.

HORNED BEAUTY

(Below) An example of the unusual "results" of evolution at this time is *Arsinoitherium*, the African counterpart of the dinocerates. This rhino-sized beast—named after the ancient Egyptian queen Arsinöe—looks superficially like a rhinoceros, but its teeth, skull, and limb bones show it to be unique. Indeed, it has its own family, the embrithopods. *Arsinoitherium* was a herbivore but not an ungulate; the latter, such as horses, walk on the adapted nails of their feet (hooves), and do not necessarily feed on plants.

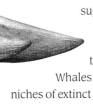

T HE APPEARANCE of whales (mammalian order Cetacea) from terrestrial ancestors is one of the great evolutionary stories, supported in recent years by fabulous new fossil finds. Looking at a streamlined dolphin or an immense filter-feeding blue whale weighing 100 tons, it is difficult to imagine how this spectacular group of living mammals evolved from terrestrial ancestors, yet the fossil evidence for this remarkable transition is (perhaps surprisingly) very strong.

Fossil evidence indicates that whales evolved from land ancestors.

The disappearance of giant Mesozoic marine reptiles near the end of the Cretaceous period had created a series of evolutionary vacuua or ecological voids, which were filled by bony fish that subsequently attained a somewhat larger size. These, however, bear no relation to the whales. Unlike the situation on land, where mammals had very rapidly occupied the niches opened by the disappearance of the dinosaurs, the time lag between the disappearance of marine reptiles and the origin of whales is in the order of at least 10 to 15 million years, suggesting that a significant and gradual process of evolutionary transformation had to be undertaken before the new niche could be colonized.

Whales did not simply fill the vacant ecological niches of extinct Mesozoic marine reptiles.

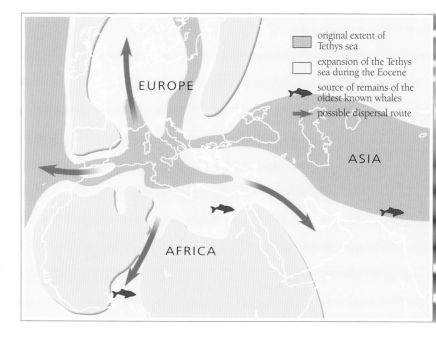

Vestigial hind-limb

TOOTHED WHALES

Toothed whales of today's oceans resemble the ancient archaeocete whales of the Paleogene more than do modern baleen whales. The latter have become extremely specialized to feed on plankton, while toothed whales probably feed in a similar way to the predatory archaeocetes.

WHALE ORIGINS

During the Paleogene, the Tethys Sea ran in an east-west direction across what is now Eurasia. Along the shores of this ancient subtropical ocean lived the ancestors of whales. These were mesonychid condylarths and they probably resembled a mixture of bear and hyena. (Bears feed happily in water and hyenas scavenge along the coast of Southern Africa). The sediments deposited along the shores of the Tethys document the successive stages in whale origins as fossils show adaptations to land, semi-aquatic, and fully aquatic ways of life.

A plausible scenario of whales' origination suggests that terrestrial mesonychids such as the huge *Pachyaena*, and even proto-whales (archaeocetes) such as *Ambulocetus*, may have lived as seashore scavengers, much like the modern brown hyena (also called the Strand Wolf) of South Africa, which survives by scavenging animal remains and small marine animals from the Atlantic beaches (strands) of southern Africa. Over a period of time during the Paleogene, it may have been that competition from purely terrestrial mesonychids such as *Pachyaena* pushed their own descendants—the ancestors of archaeocetes—to live progressively more and more in the nearshore marine realm, from which the menace of huge marine predators had recently disappeared. The evolution of limbs that were adequate for swimming would have directed the evolution of archaeocete ancestors as they became more adept at living off marine animals. Finally the link with the terrestrial realm was broken and ancient cetaceans became fully marine.

It is useful to consider that living bears, especially polar bears, are proficient swimmers, and that seals, sea lions, and walruses arose from bearlike ancestors in the Oligocene–Miocene. In the light of such functional analogues, the terrestrial model of whale origins going back to the mesonychids seems entirely plausible.

THE EARLY evolution of whales is illustrated by partial skulls and skeletons of about five different types that range in age from early to mid-Eocene times. All of these animals were found in near-shore marine sediments in modern Pakistan, where, 50 million years ago, the northern shores of the ancient Tethys Seaway lay. These early proto-whales of Pakistan's Eocene are referred to as archaeocetes and were very different from modern whales in their anatomy and lifestyle.

In the northern reaches of the ancient Tethys Seaway (now in Pakistan), ancient whales flourished 50 million years ago.

Cetaceans certainly arose from mesonychid condylarths, the abundant archaic mammals of the Paleocene and Eocene. The broad-bladed, cusped zeuglodont teeth give the clue to this, as do the extraordinary anatomical changes from mesonychid to whale: body, limbs, ears, and progressive telescoping of the skull.

Ambulocetus ("walking whale") from the Middle Eocene—the earliest marine archaeocete known to date—had long femurs (thigh bones) and long paddle-like feet. It is interpreted as an extremely large otter-like or seal-like, foot-propelled, swimming animal that was about 11ft (3.5m) in length with a powerful, 2ft (0.6m) long skull. The limbs were adapted for swimming, though *Ambulocetus* was clearly mobile on land. It probably adopted a crouched posture and hauled itself around like an unusually agile seal. *Indocetus,* from about the same time, was similar, but its fused pelvic-backbone connection and tail structure show it to be more of a tail-propelled swimmer. All subsequent cetaceans were tail-swimmers, not paddlers.

A more recent find of a similar animal, *Rodhocetus*, casts new light on whale origins. The skull is long, powerful, and armed with the immense zeuglodont teeth of other archaeocetes, but its pelvis and backbone are much better preserved. It had very high spines, or processes, on its backbones and a pelvis that articulated directly with the backbone. It would appear that *Rodhocetus* supported its weight on land, and yet, at the same time, its neck was short, the thigh bone was small, and the pelvic backbones were free to move. These details show that this animal was intermediate between foot and tail-propelled swimmers. It could both swim and move well on land.

original extent of Tethys sea

expansion of the Tethys sea during the Eocene

source of remains of the oldest known whales

possible dispersal route

EUROPE

ASIA

AFRICA

The archaeocete whale *Basilosaurus*, **looking like a classic sea-serpent, was fully adapted to a marine existence.**

Later Eocene whales had similar skulls—and, on the basis of their skull anatomy, similar predatory tendencies—to *Rodhocetus* and *Pakicetus*, but evolved long serpentine bodies up to 40ft (12m) in length. These were the basilosaurids, and they were the first really large whales to appear in the seas. The hindlimbs are still present, with all the elements still in place, though much reduced in size and of no use for swimming.

The first fully marine archaeocetes, such as *Protocetus*, had bodies that were probably less than 10ft (3m) long. Its features are echoed in the skulls of its evolutionary cousin *Pakicetus,* which exhibits a skull that is about 1ft (0.3m) long, with jaws that are also long, but not slim as those in dolphins. The jaws were very stout and

the teeth that filled them were extremely impressive: huge triangular teeth reminiscent of the teeth of the modern great white shark, but less sharp and more robust, with a series of tough cusps cascading down the front and back edges. This tooth form is very characteristic of archaeocetes and is referred to as zeuglodont for its broad serrated shape. Both *Pakicetus* and *Protocetus* possessed this dentition.

Protocetus had a sinuous body with strongly reduced fore and hind limbs, but the hind limbs were still significant elements and the forelimbs were almost certainly sufficiently strong to allow some limited locomotion on land—in great contrast to modern whales. Recent finds in Pakistan have revealed quite extraordinary animals that clearly represent intermediates in the functional stages of cetacean evolution.

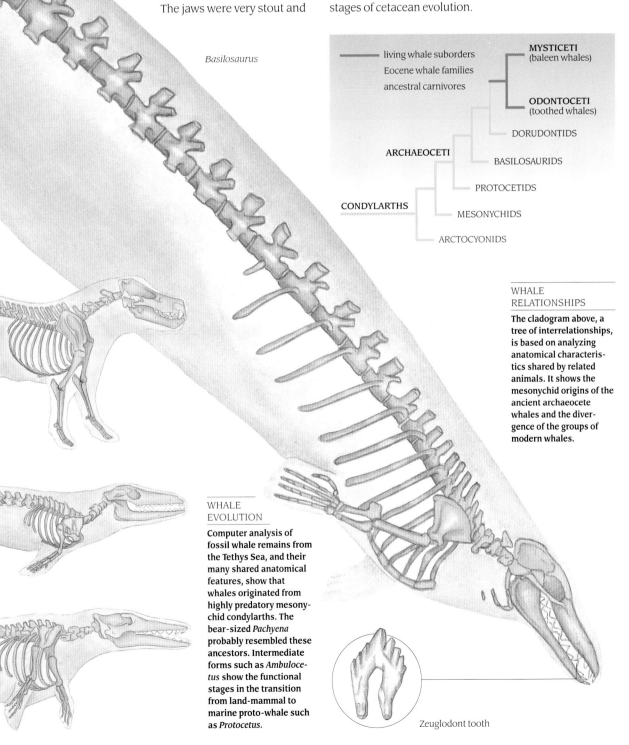

Basilosaurus

Pachyaena

Ambulocetus

WHALE EVOLUTION

Computer analysis of fossil whale remains from the Tethys Sea, and their many shared anatomical features, show that whales originated from highly predatory mesonychid condylarths. The bear-sized *Pachyena* **probably resembled these ancestors. Intermediate forms such as** *Ambulocetus* **show the functional stages in the transition from land-mammal to marine proto-whale such as** *Protocetus*.

Protocetus

— living whale suborders
  Eocene whale families
  ancestral carnivores

MYSTICETI
(baleen whales)

ODONTOCETI
(toothed whales)

DORUDONTIDS

ARCHAEOCETI

BASILOSAURIDS

PROTOCETIDS

CONDYLARTHS

MESONYCHIDS

ARCTOCYONIDS

WHALE RELATIONSHIPS

The cladogram above, a tree of interrelationships, is based on analyzing anatomical characteristics shared by related animals. It shows the mesonychid origins of the ancient archaeocete whales and the divergence of the groups of modern whales.

Zeuglodont tooth

PALEOGENE

Among the many sites in the world where fossils are found in abundance, a small number stand out. These precious sites include Lake Messel near Frankfurt, Germany, which is the source of a collection of Early Eocene fossils, about 50 million years old. In many ways the preservation of the animals from Lake Messel is perhaps the most perfect and extraordinary in the world of paleontology. Here are fossils found and preserved as nowhere else: hair, feathers, membranes, stomach contents, and even internal organs have remained intact in some cases.

A small, deep lake in what is now southern Germany provided perfect conditions for preserving a great variety of Eocene species.

The surrounding geology of the Messel lake site shows that it was a small, relatively deep lake in a fault valley with anoxic (non-oxygenated) bottom waters. Paleoenvironmental indicators, such as the types of fossils there, show that it stood in the middle of a lush tropical forest. Examples of these indicators include the plant fossils such as laurel, oak, beech, citrus fruits, vines, palms, rare conifers, and water lilies. The geological strata are mostly bituminous (organically derived tar) clays, which must have accumulated in the warm subtropical water.

The lake was unstratified, but its sediments reveal that conditions at the bottom were extremely unfavorable to life. Consequently, if the carcass of a dead animal drifted down through the warm waters to rest on the lake bed, there were no scavengers around to consume it. Poor oxygen content and extremely quiet conditions allowed the clays to embalm the body in a fine shroud of sediments. So many animals were preserved in these sediments that the carbon from their bodies formed the oil

MAMMALS OF MESSEL

Thirty-five different species of mammal have been found at Messel, including bats, insectivores, carnivores, ungulates, anteaters, primates, marsupials, pangolins, and rodents. There were oddities like *Leptictidium* that resemble no living animal, and creatures such as *Eurotamandua*, which is nearly identical to modern anteaters.

1 *Archaeonycteris* (the first known bat)
2 *Messelobunodon* (a primitive artiodactyl)
3 *Propalaeotherium* (a primitive horse)
4 *Leptictidium* (an insectivore)
5 *Paroodectes* (a miacid)
6 *Eurotamandua* (an anteater)
7 *Pholidocercus* (a hedgehog)
8 *Chelotriton* (a terrestrial salamander)

deposits that made the site valuable for industry in the nineteenth century and beyond.

The Messel deposits preserve insects and other invertebrates, vertebrates, and plants that represent input from nearby terrestrial settings. Among the non-aquatic fossils, air-transported ones such as leaves, pollens, bats, birds, and insects predominate. Most of the small amphibians that are found here are concentrated near the mouth of a shifting drainage that fed the lake. These are critical considerations, for the remains of animals that lived in forests are exceedingly rare in the fossil record (such as the earliest cats, which probably lived in forests, then as now); this is because of forest conditions: an animal that dies and drops to the forest floor is rapidly eaten by the numerous scavengers that are present in the forest. The roots and shoots of trees and shrubs also disturb the forest floor, making it a mechanically active and destructive environment in which bones are quickly destroyed. In addition, forest floors are highly acidic and

EXCEPTIONAL PRESERVATION

The amazing preservation of the fossils from the site of Lake Messel allows unique insights into the anatomy, relationships, and ways of life of these long-dead species. Many of these are small forms that lived in or around trees—an environment that usually does not preserve such tiny remains; here though, the animals have been deposited in the still, anoxic waters of a deep forest lake and their delicate structures are found in perfect order, as in the bat *Archaeonycteris* (far left), and in many species of land birds.

dissolve bones readily. One aspect of major significance about the fossil site at Lake Messel is that it preserves a range of hitherto unknown forest animals from the European Eocene of 50 million years ago.

Two particular groups of animals at the Messel site have attracted the most attention from paleobiologists: terrestrial mammals and bats. A number of the mammals found at Messel are unique to this locality and provide critical information into mammalian history. The oldest anteater, *Eurotamandua*, comes from this site and is very like living anteaters except that it still has cheekbones, which have been lost in modern anteaters. *Eurotamandua*'s presence in Europe is a mystery, because all living anteaters are found only in South America. A similar evolutionary conundrum is *Eomanis*, the fossil pangolin from Messel, which belongs to a mammalian group found only in Africa and southeast Asia. It may be that both anteaters and pangolins originated in Europe despite now being absent from that continent.

Other terrestrial mammals from Messel include more unusual forms whose affinities with modern living mammals is often obscure. The early horse-like animal *Propalaeotherium* is only the size of a spaniel and shows the primitive arrangement of its feet with the presence of four little hooves in each foot. Its tooth crowns were low, showing that it fed on soft leaves and tropical fruits in the forests; it was a browsing animal and its affinities are not difficult to establish. The same cannot be said for the small bipedal insectivore *Leptictidium*, a mammal that stood only eight inches high. It had a short, strong trunk region, very long tail, and hind legs like a bandicoot. Its front limbs were too short to be used for walking, but instead of moving like a modern rabbit or hare, it seems to have run fast with alternating steps. This curious little creature has no analogous form among living animals. The Messel site preserves even bats, which are a great rarity in the fossil record, and even shows that one particular type fed exclusively on butterflies.

THIS might have been the scene in central Asia around the time of the Paleocene–Eocene boundary 55 million years ago. As today, much of the region is situated far from the ocean and so the ameliorating effects that the sea exerts on terrestrial climates was reduced. This resulted in seasonal extremes, with summers being ferociously hot and winters excessively cold.

Andrewsarchus might have been the largest land predator to have ever lived.

Despite this, lush forests were in abundance because of tropical global conditions. Open plains were in evidence but not yet to the extent that they were to reach from the Oligocene onwards. The forests provided huge amounts of plant food for browsing herbivorous mammals; these became commonplace and could attain immense sizes. In response to this great bounty, the scavengers and active hunters of the time assumed much larger body dimensions than modern predators.

At the edge of a subtropical forest, *Andrewsarchus*, a gigantic hooved mesonychid condylarth, scavenged a dead *Embolotherium* with its huge spike-like canines and tearing triangular premolars. Both animals were the size of a hippopotamus, with the embolothere weighing in at an estimated 3 tons. Almost equally massive predators closed in for a look, such as the *Sarkastodon*. About 10ft (3m) long, it greatly outweighed any living bear, and its slicing, creodont molars and robust, bone-cracking premolars suggest that it was a more carnivorous animal than our largely omnivorous modern bears. Another mesonychid, the lion-sized *Harpagolestes*, also approached the kill. The few areas of open plain in central Asia provided a hunting ground for packs of the hyenodont creodont *Paracynohyaenodon*. The marten-like miacid *Vulpavus*, one of a long line of "true" carnivores that would eventually replace the Paleogene giants, kept clear of the competition between these huge carnivores until they had eaten their fill, while tiny primates, a recently evolved order, remained safely in the trees.

1 *Harpagolestes* (a mesonychid)
2 *Paracynohyaenodon* (a creodont)
3 an early primate
4 *Sarkastodon* (a creodont)
5 *Andrewsarchus* (a mesonychid)
6 *Embolotherium* (a titanothere)
7 *Vulpavus* (a miacid)

THE EVOLUTION OF MAMMALS

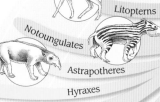

MOTHER'S MILK

A major and unique feature that helps to define a mammal is the presence of mammary glands from which the live-born young suck milk.

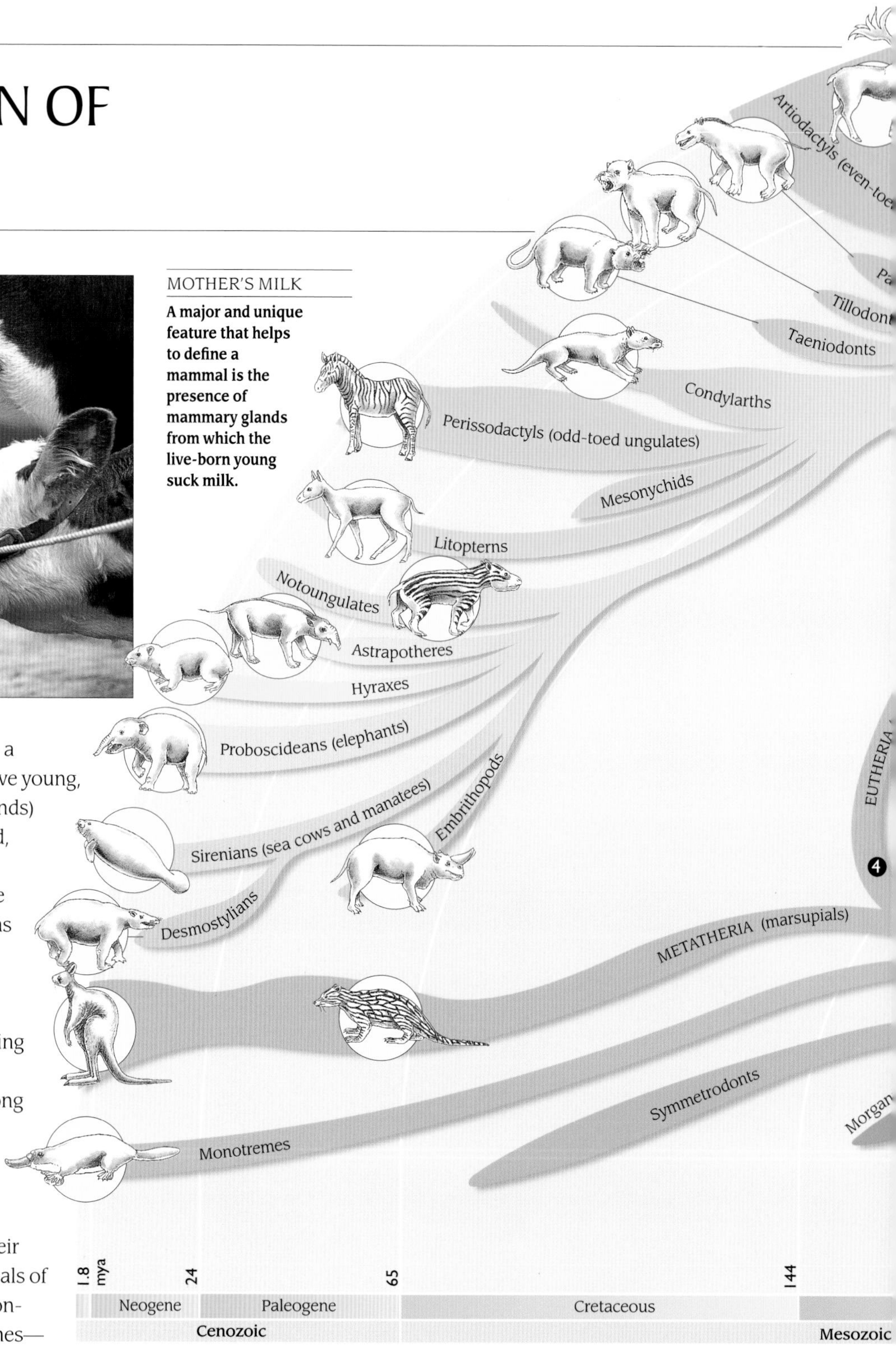

Artiodactyls (even-toed)

Pa

Tillodonts

Taeniodonts

Condylarths

Perissodactyls (odd-toed ungulates)

Mesonychids

Litopterns

Notoungulates

Astrapotheres

Hyraxes

Proboscideans (elephants)

Sirenians (sea cows and manatees)

Embrithopods

Desmostylians

EUTHERIA

METATHERIA (marsupials)

Symmetrodonts

Morgan

Monotremes

THE CHARACTERISTICS used to define a mammal (warm-blooded; giving birth to live young, which they suckle on milk [mammary] glands) are no use for interpreting the fossil record, since such features are not preserved. However, there are many clues that may be read from tough fossilized bones, almost as easily as from a recent dry skull.

Cladistic analysis—which looks at the recentness of common ancestry of animal groups and their relationships, by comparing anatomical data—demonstrates that the probable origins of true mammals lay among the mammal-like reptiles, the shrew-sized trithelodont cynodonts of the Middle Triassic. Certainly these little animals show some remarkable resemblances to mammals, especially in the structure of their skulls. Cladistic analysis of the tiny mammals of the Mesozoic era, however, is extremely contentious; for example, the living monotremes—egg-laying mammals of Australia—have much more primitive anatomy than many extinct Mesozoic mammal groups, such as the long-lasting multituberculates. One fact beyond doubt is that the living marsupials and placental mammals are the most advanced of all—that is, they have the most derived characteristics.

When cladistic analysis is applied to fossil mammals of the Cenozoic, certain relationships emerge. For instance, the anteaters, sloths, and armadillos are close cousins of the pangolins, while rabbits are found along with rodents and elephant shrews. Tree shrews, bats, and sugar gliders are actually rather closely related to primates, with which they all form a large group called the archontans. The ungulates (hoofed mammals) combine with the cetaceans, sirenians, hyraxes, elephants, and the aardvark to form the group Ungulata.

mya 1.8 | 24 | 65 | 144

Neogene | Paleogene | Cretaceous

Cenozoic | Mesozoic

1 225 million years ago in North America, *Adelobasileus,* a shrew-sized animal, is the oldest mammal

2 The lineage splits into a series of still primitive forms such as multituberculates

3 The monotremes originate during the Late Jurassic

4 The marsupial–placental split occurs in the Early Cretaceous of North America with *Alphadon,* a marsupial.

5 Huge radiation of mammals as dinosaurs become extinct

Analysis of the fossil mammals of the Cenozoic also produces interesting patterns. For instance, the taeniodonts, the giant gnawing "bears" of the Paleogene, seem to be descended from a group that is close to the origins of the creodonts—the huge Paleogene cousins of the Carnivores. This ancestral group is probably the arctocyonids, small mammals that are among the very earliest of the condylarths. In another interpretation, the huge dinocerates—such as *Uintatherium*— come out alongside the the cetaceans and the perissodactyls. The clustering of whales with two groups that are obviously ungulate-like suggests that the cetaceans had origins among hoofed groups as well. This is supported by the close placement of the whales to the mesonychid condylarths, the primitive hoofed predators of the Paleocene.

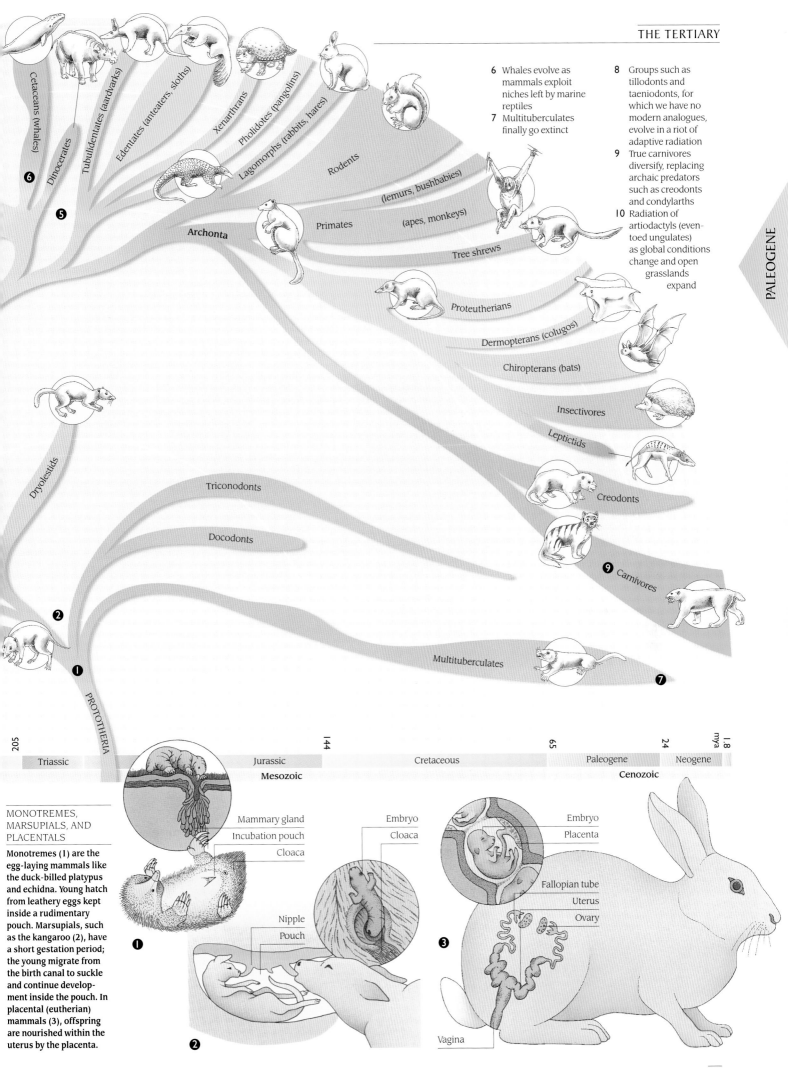

6 Whales evolve as mammals exploit niches left by marine reptiles

7 Multituberculates finally go extinct

8 Groups such as tillodonts and taeniodonts, for which we have no modern analogues, evolve in a riot of adaptive radiation

9 True carnivores diversify, replacing archaic predators such as creodonts and condylarths

10 Radiation of artiodactyls (even-toed ungulates) as global conditions change and open grasslands expand

PALEOGENE

Cetaceans (whales)
Dinocerates
Tubulidentates (aardvarks)
Edentates (anteaters, sloths)
Xenarthrans
Pholidotes (pangolins)
Lagomorphs (rabbits, hares)
Rodents
(lemurs, bushbabies)
(apes, monkeys)
Primates
Archonta
Tree shrews
Proteutherians
Dermopterans (colugos)
Chiropterans (bats)
Insectivores
Leptictids
Creodonts
Carnivores
Dryolestids
Triconodonts
Docodonts
Multituberculates
PROTOTHERIA

205 | 144 | 65 | 24 | 1.8 mya

Triassic | Jurassic | Cretaceous | Paleogene | Neogene
Mesozoic | Cenozoic

MONOTREMES, MARSUPIALS, AND PLACENTALS

Monotremes (1) are the egg-laying mammals like the duck-billed platypus and echidna. Young hatch from leathery eggs kept inside a rudimentary pouch. Marsupials, such as the kangaroo (2), have a short gestation period; the young migrate from the birth canal to suckle and continue development inside the pouch. In placental (eutherian) mammals (3), offspring are nourished within the uterus by the placenta.

Mammary gland
Incubation pouch
Cloaca

Nipple
Pouch

Embryo
Cloaca

Embryo
Placenta
Fallopian tube
Uterus
Ovary
Vagina

THE EVOLUTION OF CARNIVORES

Ecological niches defined by the roles of top predator were opened up by the demise of the dinosaurs in the late Cretaceous period. Because the initial evolutionary radiation of early Cenozoic mammals did not include any large carnivores, a number of completely different animal groups were in a position to take over.

In South America the giant "terror birds" such as *Phorusrhacos* occupied this niche throughout the Eocene and most of the Oligocene, only meeting competition later with the advent of large mammalian carnivores— not placental mammals but marsupials. The jaguar-sized *Arminiheringia* posed a serious threat at least until the very latest Oligocene. By the start of the Miocene, the phorusrhacids were close to extinction and the marsupicarnivores soon came up against placental carnivores. Modern carnivorous marsupials are mainly in Australia and include the Tasmanian devil (*Sarcophilus*), a dasyurid, but no other much larger than a cat. The last medium-size marsupial predator on the planet, the Tasmanian wolf (*Cynocephalus*), became extinct in 1933.

Elsewhere, several groups of sizeable vertebrates competed for predatory niches. In Europe, "true" carnivores (*Carnivora vera*) vied with huge archaic creodonts and with an unusual group of fully terrestrial crocodilians. The latter had a relatively short evolutionary history, perhaps caught between competition from mammalian carnivores on land and the superbly adapted semiaquatic crocodilians in the rivers and estuaries.

In Asia, marsupials remained small, as did the earliest true carnivores. The main predators belonged to the creodonts and the mesonychid condlyarths. Many were exceptionally large and anatomically unusual: *Andrewsarchus* was the largest-ever mammalian land predator/scavenger, and *Sarkastodon* was a hulking, solitary, bear-like hunter. Another large mesonychid predator was *Pachyaena*. The wolf-sized creodont *Paracynohyaenodon* had the proportions of a low-slung dog and may have been a pack hunter, unlike *Sarkastodon*. All found easy prey among the browsing mammals and herbivores of Asia's lush Paleogene forests. However, many Asian predators seemed to have been better adapted for scavenging rather than full-fledged meat eating.

As the world cooled, subtropical forests disappeared and the biomass of herbivorous mammals was reduced. For large predators, the new, smaller, quicker herbivores yielded too little meat for the energy expended by chasing prey on short legs supporting heavy bodies. Longer legs are more efficient, and this may have been a factor in the subsequent success of true carnivores. They also had the advantage of being able to use their teeth to eat roots, nuts, and fruits; creodonts, whose teeth were only shears, could not. Paradoxically, it seems that true carnivores succeeded by not eating meat all the time.

CARNIVORE GROUPS

During the Paleogene, "true" carnivores were small miacids—marten-like animals that were very different from the hulking predators that existed then. These latter forms included mesonychids, creodonts, and, in South America, the borhyenid marsupials. The fundamental carnivore split into the vulpavines ("dog branch") and viverravines ("cat branch") occurred among the miacids some 60 million years ago. Living vulpavines (Canoidea) include the dog, bear, and weasel families, some members of which are omnivorous, while the more purely carnivorous viverravines (Feloidea) divide into four fundamental groups, including the superficially dog-like hyenids. Extinct feloids include the saber-toothed cats, while a number of canoid lineages have become extinct. Among these are the bone-crushing dogs or borophagines that filled the hyenid niche in North America, the immense amphicyonids ("bear-dogs"), and the nimravids or "paleo-sabers." Nimravids are essentially saber-toothed cats, yet they are anatomically more closely related to dogs than to cats.

1 Various small meat-eating mammals, including didelphid-marsupials and condylarths, appear
2 *Cimolestes* shows the beginnings of specialized molar teeth: the carnassial shear
3 Miacids are the first "true" carnivores
4 Creodonts, an archaic carnivore group, are the top predators
5 A split into "dogs" and "cats" takes place via the miacids, *Vulpavus* and *Viverravus*
6 The Canoidea and Feloidea split into a number of lineages
7 Saber-toothed cats represent the zenith of Cenozoic carnivorousness

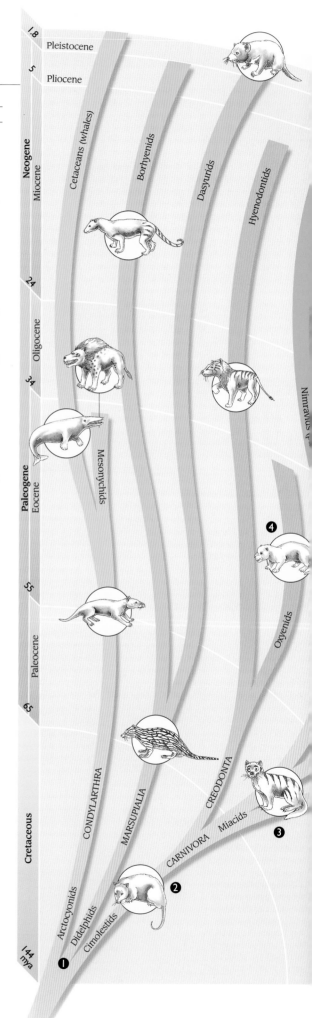

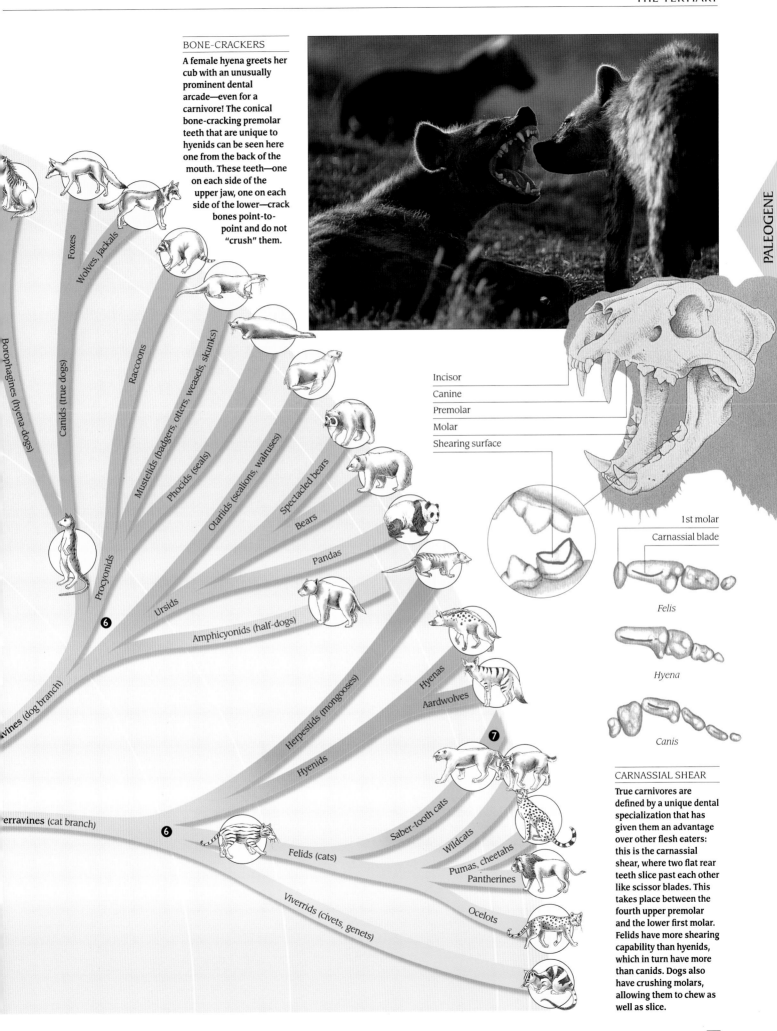

BONE-CRACKERS

A female hyena greets her cub with an unusually prominent dental arcade—even for a carnivore! The conical bone-cracking premolar teeth that are unique to hyenids can be seen here one from the back of the mouth. These teeth—one on each side of the upper jaw, one on each side of the lower—crack bones point-to-point and do not "crush" them.

Foxes

Wolves, jackals

Raccoons

Borophagines (hyena-dogs)

Canids (true dogs)

Mustelids (badgers, otters, weasels, skunks)

Phocids (seals)

Otariids (sealions, walruses)

Spectacled bears

Bears

Pandas

Procyonids

Ursids

Amphicyonids (half-dogs)

⑥

Herpestids (mongooses)

Hyenas

Aardwolves

⑦

Hyenids

...vines (dog branch)

...erravines (cat branch)

⑥

Saber-tooth cats

Wildcats

Felids (cats)

Pumas, cheetahs

Pantherines

Viverrids (civets, genets)

Ocelots

Incisor

Canine

Premolar

Molar

Shearing surface

1st molar

Carnassial blade

Felis

Hyena

Canis

CARNASSIAL SHEAR

True carnivores are defined by a unique dental specialization that has given them an advantage over other flesh eaters: this is the carnassial shear, where two flat rear teeth slice past each other like scissor blades. This takes place between the fourth upper premolar and the lower first molar. Felids have more shearing capability than hyenids, which in turn have more than canids. Dogs also have crushing molars, allowing them to chew as well as slice.

THE NEOGENE
24 – 1.8
MILLION YEARS AGO

THE NEOGENE *period, which comprises the Miocene and Pliocene epochs, extends from 24 million years to 1.8 million years ago. During this time, the world experienced three events of great significance that are unique to this period: the evolution of modern mammals, including the appearance of the first humans, and the onset of a cycle of ice ages which dominated the Quaternary period that followed. The spread of grasslands, which had begun in the Paleogene, continued, leading to the diversification of grassland fauna.*

During the Neogene, the modern groups of mammals that are familiar in today's ecosystems radiated and evolved from more archaic Paleogene ancestors. Most of the anatomically unusual Paleogene mammals went extinct, including all the large endemic land mammals of South America, and were replaced by more derived types. The modernization of the Cenozoic mammal fauna took place over a mere 24 million years— a very brief time in geological terms.

THE BOUNDARY between the Paleogene and the Neogene is not marked by any mass extinction or other historical event, but the latter period is distinguished by the occurrence of several significant biotic changes: the spread of herbaceous plants and grasses; the development of major new communities, typified by important changes in the teeth of herbivores; and the appearance of completely new families of mammals. The spread of grassland habitats was a result of climate change—the global cooling that was gradually taking place as ice sheets spread, first of all in Antarctica, and subsequently in the Arctic. Sea levels rose and fell and the temperature and humidity fluctuated, becoming cooler

During the Neogene the climate became increasingly cool and dry. Tropical forests diminished and grasses took over.

KEYWORDS

AUSTRALOPITHECINE

BALEEN

CETACEAN

GLYPTODONT

HOMINID

MARSUPIAL

MESSINIAN CRISIS

MYSTICETE

NOTOUNGULATE

PINNIPED

PRIMATE

UNGULATE

and drier overall. Local tectonic events caused parts of east Africa, western North America, and South America to receive less rainfall than they had previously enjoyed, with an impact on animal populations. New land-bridges formed that allowed the exchange of animals and plants.

The most significant ecological pattern of the Neogene, prior to the cooling that foreshadowed the Pleistocene ice ages, was the redevelopment and diversification of ecosystems that were dominated by large herbivores. Although there was an increase in the extent of grasslands from the mid-Oligocene onwards, forests have remained one of the most characteristic types of vegetation on earth, for there has not been an expansion of open terrains at the expense of closed vegetation. Nevertheless, although forests have remained significant, one of the most evident outcomes worldwide during the Neogene was the diminishing of forests and the spread of more open woodlands, grasslands, and deserts.

PALEOGENE	24 mya		20		NEOGENE		15
Series							MIOCENE
European stages	AQUITANIAN			BURDIGALIAN		LANGHIAN	SERRAVALLIAN
N. American stages		ARIKAREEAN			HEMINGFORDIAN		BARSTOVIAN
Geological events				Australian plate moves north; Indian craton collides with Eurasia; uplift of Himalayas			
	Alpine orogeny continues; major deformation of Alps and Carpathians						
Climate				Temperate boreal savannah, low rainfall			
Sea level				Moderate			
Plants				Spread of grasses and herbaceous plants adapted to dry habitats			
Animals			Diversification of whales			Diversification of rodents, songbirds, sna	

That conditions were different from today was mainly a result of glacial cycles and not through major changes in continental distribution. Furthermore, complex regional variations and temporal (time) fluctuations are all notable in the climatic record of the Neogene period. These variations would have produced effects on local biotas that may have created rapid adaptive evolution in the animals of the area. Such changes, however, are extremely difficult to disentangle from the normal level of evolution that occurs over longer periods of time. The development of open vegetation certainly drove much of the evolution of Neogene mammals, which developed features that are strongly associated with an adaptive response to such ecosystems. These include increased development of high-crowned teeth that would withstand the wearing effect of the new high-silica content grasses; the ability to run fast and other locomotor characteristics related to movements in open habitats; larger body sizes in herbivores (as a result of the need to process high volumes of low-quality fodder); diversification of small herbivores that used burrows; and the development of a wide range of carnivores to feed on different herbivore classes.

The Neogene, despite its short 24-million-year history, showed rapid and profound evolution of life in response to widely fluctuating climates. In addition to the large herbivores, there was a tremendous radiation of small animals: songbirds, frogs, rats, and mice. They fed on the seeds of the grasses, or the insects that pollinated them, and, in the case of the rodents, burrowed in the dry terrain. In pursuit of the rodents came snakes.

In many ways the floras and faunas of the Neogene were similar to those of today, especially those of the later part of the period. Nevertheless, some of the mammals that inhabited the Miocene and Pliocene worlds are remarkable for their great distinctness from our recent mammals. According to a general rule of zoology, which was noted in the nineteenth century by Charles Darwin, invertebrates evolve more slowly than vertebrates. Compared with the modest changes in the fossil record of invertebrate life, the modernization of vertebrates during the brief span of the Neogene was extreme.

THE DEVELOPMENT of ice sheets over northern hemisphere continents began about 2.4 million years ago. It is interesting to note the enormous capacity of the polar ice sheets to overcome all other climatic effects. Variations in the amplitude (magnitude of variation) of the glacial cycles that characterize the Neogene, in the development of polar ice sheets and in other factors, resulted in a climatic history so complex that it is only just being unraveled now, using isotope geochemistry and novel paleontological approaches. However, there was certainly an overall global cooling as the world continued in its "icehouse" state, which also interacted with local factors such as mountain building and rifting events. The development and establishment of very steep thermal gradients from the poles to the tropics has been one of the major climatic shifts since the Eocene.

The complex climatic history of the Neogene is only just being unfolded.

Also associated with the global cooling was a significant global drying. This was in part due to reduced evaporation from the colder oceans and in part to the expansion of the polar ice caps and the formation of deep cold ocean waters (the psychrosphere); it is a form of physiological drought, in which water is present but unavailable to animals because it is in the wrong physical state: ice.

While studies of Neogene climate usually focus on the high latitudes of the northern hemisphere, patterns of climatic change during this time—and their effects on

See Also

THE PALEOGENE: *Psychrosphere, Grasslands, Whales, Mammals*
THE PLEISTOCENE: *Glaciation, Gulf Stream, Humans*
THE HOLOCENE: *Caribbean, Andes, East African Rift valley*

THE NEOGENE

The Neogene occupies less than one percent of the latest period of Earth's history. The most significant of the changes in ecosystems during this time had to do with explosive rates of evolution and expansion among mammals, which had a profound effect on future developments.

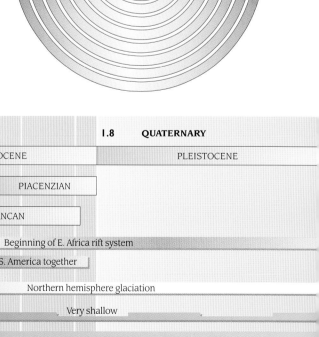

NEOGENE

10		5	1.8	QUATERNARY
		PLIOCENE		PLEISTOCENE
TORTONIAN		MESSINIAN	ZANCLEAN	PIACENZIAN
CLARENDONIAN		HEMPHILLIAN		BLANCAN

Evaporation of Mediterranean Sea | Beginning of E. Africa rift system
Panama arc joins N. and S. America together
Continued cooling, growth of Antarctic ice sheet | Northern hemisphere glaciation
Falling | Very shallow
Spread of high-silica grasses
Decline of large, odd-toed ungulates; radiation of even-toed ungulates | • First hominids

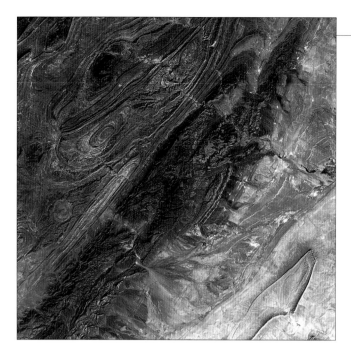

The Panamanian land-bridge that now links South America with North America via the countries of Central America was slowly forming, and the Caribbean was still a natural extension of the Pacific rather than an embayment of the Atlantic as it is now. Some volcanic islands had risen in the Caribbean region; others remained underwater, where they accumulated great layers of limestone that later formed the distinctive karst landscapes of the Greater Antilles islands. This land-bridge is significant for another reason: it blocked the westward movements of warm waters from the southern Atlantic and redirected them to the northeast, forming the Gulf Stream.

DURING the Jurassic period, prior to the collision between the European and African plates, the Tethys—an Alpine Mesozoic ocean—had formed as an extremely long rift basin, because of the opening of the Atlantic Ocean and the eastward movement of Africa in relation to Europe. Closure of the Tethys began late in the Cretaceous period as the African plate continued to move. At the time of the Paleogene–Neogene boundary, the Tethys closed still further. What remained formed the modern Mediterranean Sea, with the Parathetys in existence to the east. Gradually, the volume of the Mediterranean Sea became much smaller as a result of the expansion of Antarctic ice sheets in the southern hemisphere, which lowered sea levels by as much as 165ft (50m) and cut off the Mediterranean from the Atlantic Ocean, causing the Mediterranean to dry up altogether at the end of the Neogene.

The African plate hit Europe early in the Cenozoic era and is still continuing to shear in an eastward direction. The Mediterranean and North Africa were directly affected as a result of the fall and rise of sea levels and the movement of the African plate. The Atlas, Alps, Pyrenees, and Carpathians are all mountains that were thrown up as a result of this collision. Several microplates caught up in the zone between the African and European

> *Mountain-building was far from finished in Europe, but changes in the seas were even more dramatic as the ancient Tethys closed.*

biotas—appear to have been equally complex in tropical latitudes. During phases of glacial aridity, the rainforests of central Africa and the Amazon basin shrank to a few isolated refuges, whereas those in Sumatra remained intact. Many low latitude regions experienced aridity when cold (glacial) periods occurred in higher latitudes, and moist conditions (as shown by high lake levels) appear to have coincided with interglacials. Disruption of the Southeast Asian rainforests also occurred repeatedly; this was due to high sea levels during the interglacial periods. The overall conclusion that can be drawn from the many independent studies of Neogene environments is that climates in the tropical regions of the globe were unstable, with a corresponding effect on the distribution and turnover of their biotas.

A GEOGRAPHICAL map of the Neogene world from about 6 million years ago—in the Late Miocene epoch—would look quite similar to that of today, with just a few major geographical differences. In the western hemisphere, the Rocky Mountains were rising in the western interior of the continent, on a wide expanse of uplift that was a leftover of the earlier Laramide orogeny in the Paleogene. Similarly, the modern Appalachians formed over the worn stumps of the ancient eastern ranges, relics of the Late Paleozoic collision of Laurentia with Baltica and then Gondwana. Continuing tectonic activity produced extensive flooding by the sea in southern California, while volcanoes formed the Cascade Range, and igneous rocks were extruded in a huge belt from Oregon all through Utah and into Arizona. The great Columbia River basalts are part of this formation. Mountain-building continued along the western edge of the Americas as the various Pacific plates pushed against the continents, raising the Andes range.

> *North America's Rocky Mountains were formed during this period, along with the ongoing rise of the steep Andes of South America.*

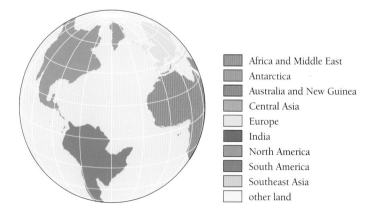

PACIFIC OCEA

Africa and Middle East
Antarctica
Australia and New Guinea
Central Asia
Europe
India
North America
South America
Southeast Asia
other land

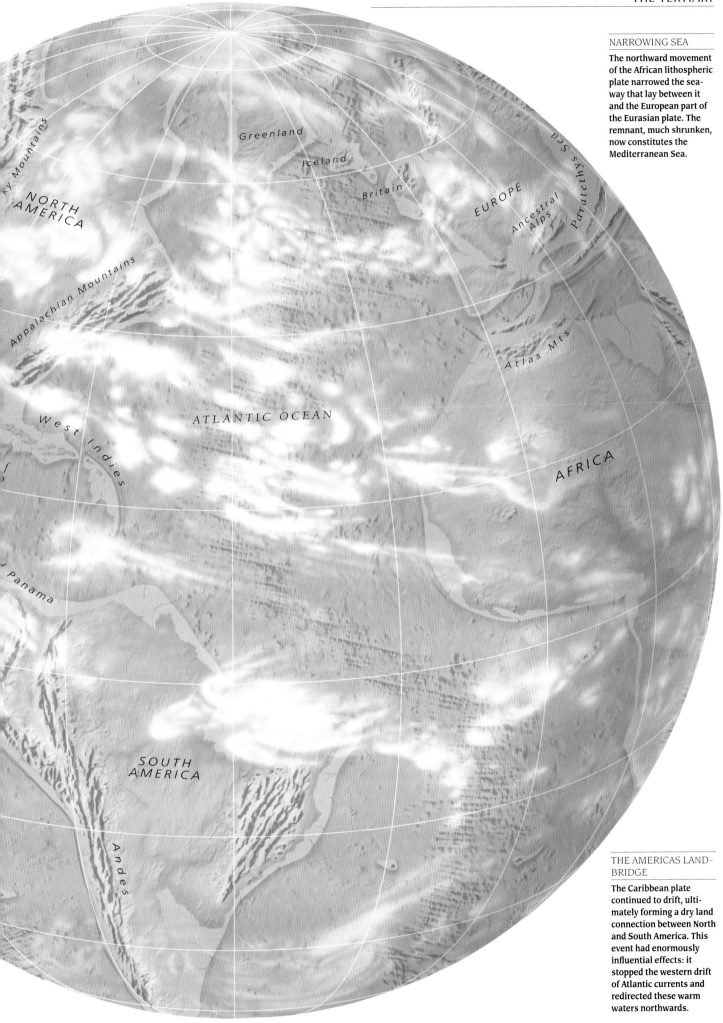

NORTH
AMERICA

ky Mountains

Appalachian Mountains

West Indies

f Panama

SOUTH
AMERICA

Andes

ATLANTIC OCEAN

Greenland

Iceland

Britain

EUROPE

Ancestral
Alps

Paratethys Sea

Atlas Mts

AFRICA

NEOGENE

NARROWING SEA

The northward movement of the African lithospheric plate narrowed the seaway that lay between it and the European part of the Eurasian plate. The remnant, much shrunken, now constitutes the Mediterranean Sea.

THE AMERICAS LAND-BRIDGE

The Caribbean plate continued to drift, ultimately forming a dry land connection between North and South America. This event had enormously influential effects: it stopped the western drift of Atlantic currents and redirected these warm waters northwards.

plate became the Italian and Iberian peninsulas, and the islands of Corsica, Sicily, and Sardinia. Approximately 155mi (250km) thickness of European lithosphere was involved in building the Western Alps; deriving from what was mainly the upper elements of the Earth's crust. The shape and distribution of these mountains mark the deformations wrought by the impact. The west-to-east curve of the mountains can be seen clearly in aerial or satellite photographs. The process continues today, as these mountains are even now being bent into a vast S shape across North Africa, Europe, and Eurasia.

This lithospheric movement (of Africa) drove the Iberian plate and southern parts of Europe to the east. Spain and Portugal—the Iberian microplate—were wrenched out of what is now the Bay of Biscay, swung around, and thrown against western Europe. Crustal deformation related to the convergence between these plates stopped during the Miocene epoch.

The movement of the African plate caused large scale geological stretching in an east-west direction, forming areas of tensional strain throughout much of the European plate, with enormous shallow cracks forming more or less at right angles to the direction of tension. Because of the direction of shear, these cracks are oriented in a north-south direction visible in satellite photographs. They include the Rhône Valley and the Rhine Gorge.

The first signs of compressional tectonics in the region of impact between the Iberian plate and the European plate actually occurred about 75 million years ago in the Late Cretaceous. The transtensional basins—such as the Bay of Biscay—caused by these motions in the Cenozoic era were transformed into foreland basins as uplift occurred in the region of the Pyrenees. Most of these basins are now landlocked in northern Spain and southern France. This did not happen to the Bay of Biscay, which remains an offshore ocean basin on the western European continental shelf. The impact caused the crumpling up of the region between the Iberian plate and western Europe to form the Pyrenees Mountains.

To THE east, the Indian continental plate was in the process of colliding with Eurasia, resulting in the uplift of the Himalayas. This process was not yet complete and the mountains were, as yet, only a low-lying range. However, their presence had a profound effect on the Asian continent as their great rivers, such as the Ganges, began to form in the Miocene. Farther north, the ancient inland Obik Sea and Turgai Strait, which had at times provided marine access between the Tethys and the northern ocean, were both now closed, but shallow seas covered large expanses of south and east Asia. The Arabian peninsula was all but an island, and Madagascar floated farther off the east coast of Africa, where rifts were just beginning to appear in the land as tectonic forces caused the continent to arch up by nearly 10,000ft (3000m). Australia was slowly moving north to its present position and was tectonically quiet, unlike the rest of the region, where volcanic islands were regularly emerging. Antarctica was surprisingly temperate at the beginning of the Miocene, with the great ice sheets only beginning to form between 15 and 10 million years ago.

The globe acquired two dramatic new features: the Himalayan mountains and the Late Miocene ice sheets that still cover Antarctica.

As the South Pole froze, the rest of the southern hemisphere was affected. There is considerable geological evidence that climates became cooler here between ten and five million years ago. Siliceous rather than calcareous sediments were abruptly deposited over larger areas of the southern oceans than before. The success of diatoms—microscopic algae with a silica "shell"—which produced these characteristic sediments indicate that upwelling of deep, cold ocean waters had increased, apparently because of steepening temperature gradients. This cooling was not a world-wide event at this time, but it had major consequences: as water became locked up in the great Antarctic ice caps, sea levels dropped to such a degree that Atlantic waters could not replenish the Mediterranean, and so it gradually dried out.

As the Pliocene epoch got underway, sea levels rose again, so that the seas stood well above the present level between about 4.5 and 3.5 million years ago. This sea level rise (transgression) left large volumes of marine deposits inland of the coastlines of areas such as California and eastern North America. Countries along the margins of the Mediterranean show some similar deposits from this period, as do the European countries that border the North Sea; Great Britain and Denmark are particularly good areas for the study of Pliocene marine sediments.

During this time of high sea stands, the environment of northern Europe was more equable than it is today. Evidence for this comes from the fossil faunas found here, especially the pollen record. Pollen analysis indicates that southeastern England was subtropical, or nearly so, which is warmer than the present climate. This is despite the fact that the world was—and still is—in "icehouse" mode. This pleasant climate came to an end with the start of the modern cycle of ice ages, a little more than three million years ago.

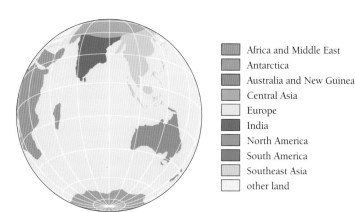

Africa and Middle East
Antarctica
Australia and New Guinea
Central Asia
Europe
India
North America
South America
Southeast Asia
other land

ASIA

Himalayas

INDIA

JAPAN

Ninetyeast Ridge

EAST INDIES

INDIAN
OCEAN

AUSTRALIA

NEW ZEALAND

ANTARCTICA

THE HIMALAYAS FORM

India was once an island; its movements are connected to Australia's. When the Indian plate collided with the southern margin of Asia, the vast upwarping of the crust beneath the two plates formed the Himalayas.

NEOGENE

AUSTRALIA AND THE ANTARCTIC CURRENT

Australia's northward drift caused the formation of a circumpolar current around Antarctica, cooling much of the world. The same drift meant that Australia moved through a series of latitudes and it has experienced contrasting climates throughout the past 30 million years.

THE FORMATION of the Himalayas began about 80 million years ago, late in the Cretaceous, when the Australian plate, carrying India, began to move north. It did not fully collide with the Eurasian plate until about 20 million years ago. The collision did not stop even there: the plate that carried India continued to push into Asia, eventually penetrating inland by 1500mi (2500km). The violence of this impact is probably unique in the long history of continental collision. It thrust up the high plateau of Tibet and pushed China and Mongolia to the east, creating a series of secondary mountains along the way.

The Himalayas began to form about 80 million years ago. It took 60 million years for the 1500-mile-long wedge of continent to push as far into the heart of Asia as we see it today.

The Himalayan front rises suddenly from the flat Ganges Plain. An effect called isostasy means that the Earth's crust under the Himalayas is very thick indeed. The crust is part of the lithosphere—the stiff outer layer of the planet—that floats on the dense partly molten rocks of the asthenosphere. As mountains are raised, their buoyancy on the surface of the asthenosphere is balanced by the formation of equally deep roots extending far down into the Earth. The young age of the Himalayas explains their great height: they have not had time to erode. Mount Everest, the world's tallest mountain, is about 27,700ft (8848m) high, and is still growing as India continues to crunch slowly into Eurasia. Even the Tibetan Plateau is 16,000ft (5000m) above sea level. At these high altitudes can be found segments of ancient seafloor called ophiolites. These are found in the Himalayas along with the remains of island-arc volcanics. Such characteristic geological features provide clues to the ancestry of this immense range of mountains: it appears that the ophiolites and island arcs were attached to the Indian craton shortly after the start of the Cenozoic era 65 million years ago, when the Indian craton was still far from Eurasia; thus the small craton must have collided with an island arc before it impacted with Eurasia.

The first point of impact was the northeast corner of India, which struck southeast Asia, slowing its progress, before the rest of it rotated to strike the southern margin of Eurasia. The Indian craton did not make contact with Eurasia near its present position until about 15 million years ago, when the oldest molasse deposits are dated. Much of the Himalayan orogeny has only taken place in the last 15 million years. These molasse deposits, because of the huge size of the Himalayan chain, cover immense areas of land. Much of the vast Ganges and Indus river deltas are located on molasse deposits, and it was the meltwater from the snow-capped peaks of the

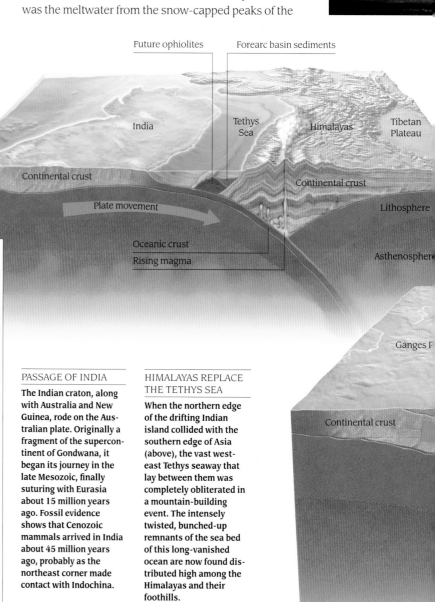

Future ophiolites   Forearc basin sediments

India   Tethys Sea   Himalayas   Tibetan Plateau

Continental crust   Continental crust

Plate movement   Lithosphere

Oceanic crust   Asthenosphere

Rising magma

Ganges P

Continental crust

relative location of the Indian landmass

- present
- 10 million years ago
- 50 million years ago
- 70 million years ago
- land uplifted by collision of India and Asia

TIEN SHAN   ASIA

KUNLUN SHAN   Tibetan Plateau   HIMALAYAS

ARABIA   INDIA

INDIAN OCEAN

PASSAGE OF INDIA

The Indian craton, along with Australia and New Guinea, rode on the Australian plate. Originally a fragment of the supercontinent of Gondwana, it began its journey in the late Mesozoic, finally suturing with Eurasia about 15 million years ago. Fossil evidence shows that Cenozoic mammals arrived in India about 45 million years ago, probably as the northeast corner made contact with Indochina.

HIMALAYAS REPLACE THE TETHYS SEA

When the northern edge of the drifting Indian island collided with the southern edge of Asia (above), the vast west-east Tethys seaway that lay between them was completely obliterated in a mountain-building event. The intensely twisted, bunched-up remnants of the sea bed of this long-vanished ocean are now found distributed high among the Himalayas and their foothills.

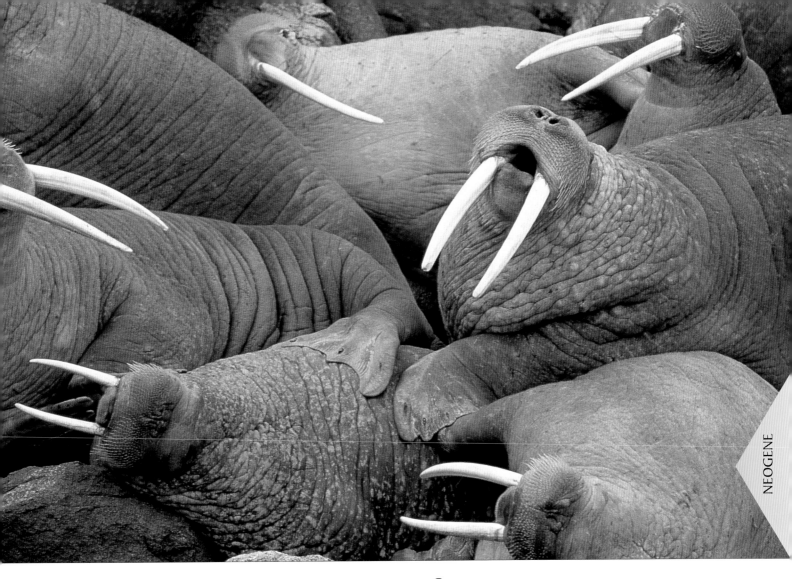

whales in the Micocene is represented by animals such as *Pelocetus*. Baleen whales strain plankton using the hair-like mats of their baleen plates, which are in fact highly modified teeth, despite the commercial name "whalebone." The thickness and number of plates varies in species. Baleen does not fossilize, however; in living mysticetes such as the gray whale this material requires considerable body reserves in order to stay functional. Consequently, the skull bones of living mysticetes show a characteristic pattern of grooves where huge blood vessels were located that provided sufficient nutrients for the baleen. The same pattern of blood vessels is found in the skull of *Pelocetus*, confirming it as a mysticete.

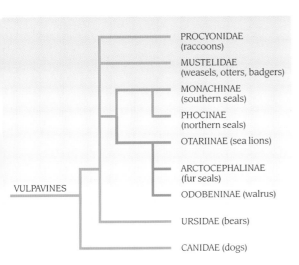

PROCYONIDAE
(raccoons)

MUSTELIDAE
(weasels, otters, badgers)

MONACHINAE
(southern seals)

PHOCINAE
(northern seals)

OTARIINAE (sea lions)

ARCTOCEPHALINAE
(fur seals)

ODOBENINAE (walrus)

URSIDAE (bears)

CANIDAE (dogs)

VULPAVINES

PINNIPED ANCESTORS

The ancestry of seals, sea lions and walruses has been intensely debated. Paleontologists originally viewed pinnipeds as diphyletic: two subsets derived from separate ancestors. Zoologists, however, normally view them as monophyletic: a natural group stemming from a single shared ancestor. This latter view is supported by molecular data. Anatomical evidence used by paleontologists suggested that sea lions originated from a bear-dog ancestor via the enaliarctines, and that seals came from an otterlike member of the mustelids. Recent fossil interpretations support a mono-phyletic origin for pinnipeds, arising from the vulpavine (dog branch) carnivores via the Ursidae (bears and relatives).

related Carnivores

suborder PHOCOIDEA

suborder OTARIOIDEA

S<small>EALS</small>, sea lions, and walruses, a group traditionally known to zoologists as the pinnipeds ("wing-footed" mammals), are another wonderful product of Neogene marine diversification. Their origin is currently a matter of considerable debate: fossils show origins from bear-dog-like mammals and their intermediate forms. Paleobiologists have identified fossils that suggest that seals and sea lions are from different lineages that converged on a very similar body pattern as they adapted to life in similar environments.

Seals and sea lions appear to have evolved separately but developed the same adaptations.

The oldest known pinniped is from Early Miocene rocks in North America. *Enaliarctos* shows anatomical affinities with the ancestors of bears, such as well-defined slicing carnassial teeth—very unlike the smaller cone-shaped teeth of modern fish-eating sea lions. It had fully developed flippers, although these were not as flattened as in modern pinnipeds. The early sea lion *Thalassoleon* comes from Late Miocene rocks in North America and is more like a modern sea lion in many anatomical details, such as the undifferentiated, single-cusped conical teeth characteristic of living pinnipeds.

The walruses also have their earliest representative in the Miocene, an animal known as *Imagotaria*. It had yet to evolve the massive tusks and foreshortened skulls of living walruses, but there are many indications of the

NEOGENE

relationship, such as the very simplified cheek teeth. Sea lions and walruses appear to have had their origins in the Pacific basin, whereas the seals seem to have originated in the Atlantic–Mediterranean region. The enaliarctines were restricted to the Miocene, and perhaps were unsuccessful in the long term due to competition from the more specialized seals and sea lions.

Along with the enaliarctines were the marine desmostylians, which were also restricted to the Miocene. Desmostylians were originally known from strange columnar teeth linked like a series of chains. Skeletal remains eventually revealed the nature of these animals, which belong to their own zoological group and have left no descendants. *Palaeoparadoxia* is a well-chosen name for an animal with a shovel-headed skull with long teeth projecting outwards from the tip of the lower jaw and a body that looks like a mixture of seal and elephant with huge splayed feet and crouched limbs. The sleek seals, sea lions, and walruses had an obvious advantage.

AUSTRALIAN HERBIVORES

A view of Australia during the mid-Miocene depicts the heavily-bodied herbivorous marsupials *Neohelos* and *Palorchestes*. Both were diprotodonts with well-developed molars and expanded front incisors. *Neohelos* was a cow-sized browser, analagous to the large African herbivores; abundant fossils suggest that it lived in herds. *Palorchestes* had a higher skull but a very low snout, which suggests that it had a tapir-like proboscis.

1 *Wakeleo*
2 *Neohelos*
3 *Palorchestes*
4 Bettong (*bettongia moyesi*)

AUSTRALIAN fauna during the Neogene period was mostly composed of marsupials. Australian marsupials are more varied and diverse than their South American counterparts simply because they have always been the major mammalian group on this continent. Their evolutionary and faunal history is closely interlinked with that of plate tectonics, migrations, and geographic isolation. The separation of Australia from Antarctica was initiated at the very base of the Cenozoic, about 60 million years ago during the Paleogene epoch. As Australia moved northwards, so did New Zealand and New Guinea, both of which are part of the same lithospheric plate. During the Eocene, Australia broke away from Antarctica and effectively isolated the marsupials that were on this landmass. Their isolation has been the critical factor in the direction of their evolution, and—with no outside influences from migrating invaders—they have evolved very differently from the rest of the world. In this way, island Australia was like Noah's legendary ark.

> *Characteristic Australian faunas, dominated by marsupial mammals and terrestrial lizards, were the result of geographic isolation.*

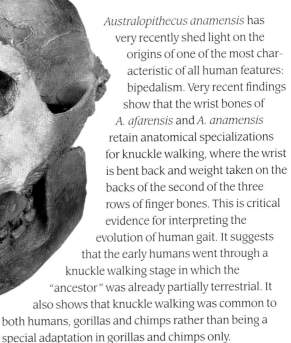

HEAVY-DUTY SKULL

Within the australopithecines there are two apparent evolutionary trends. Some individuals are termed "gracile", with light bones and modest teeth and jaws, while others are "robust". The massive cranium of this robust individual is characterized by exceptional jaw power. It is a strongly-buttressed face with flat cheeks, strong cheekbones, and a bony ridge (sagittal crest) on top of the skull. The gracile line gave rise eventually to *Homo*, but the robust forms disappeared without descendants.

FIRST TOOLS

These tools (below right) from Olduvai Gorge are grouped under the name Oldowan Industrial Complex. Dating from around 2.4 million years ago, they represent the dawn of human technology. They are simply flaked cores, made from pebbles or chunks of rock, which were used for chopping and scraping.

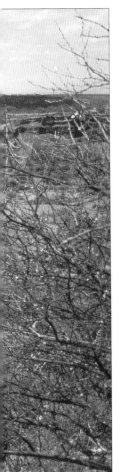

The line leading to modern humans includes as many as twelve species of *Australopithecus*. Until 1990, the australopithecines had been assigned to one genus, *Australopithecus*, but since then, new finds suggest that there are in fact three genera: *Ardipithecus,* the oldest known human at 4.4 million years old; *Australopithecus* (the gracile australopithecines); and *Paranthropus* (the robust ones). Regarding *Homo*, current views indicate perhaps six species (or seven, if the Neandertals are assigned their own branch of the evolutionary tree), and the search for the oldest human being is in many ways like the holy grail of fossil studies. Nevertheless, *Ardipithecus* is currently the oldest known human fossil; it has relatively large canine teeth, narrow molars, and thin enamel, indicating a diet of leaves and fruit; these teeth are more hominine than any of the living great apes. As for the australopithecines, they may be viewed as the sister group to *Homo*, a situation confirmed by cladistic analysis. Another ancient hominine termed *Australopithecus anamensis* appears to be intermediate between *Ardipithecus* and later species and is a specimen in which the lower leg bones suggest that it was bipedal.

HOWEVER, the most substantial remains of fossil hominids were the many skeletons belonging to a species named *Australopithecus afarensis* ("Southern ape from the Afar basin"), one of which was discovered by Donald Johanson at Hadar in Ethiopia in 1974. It was given the nickname "Lucy " and is possibly the most famous of our human ancestors. The skeleton of Lucy was 40 percent complete—an extraordinary amount of preservation for such fossils. She lived about 3.18 million years ago and was about twenty when she died. She walked on two legs, but with slightly bent limbs. Analysis of fossil pollen and animal bones indicate that her environment ranged from open grassland to woodland. Individuals of *A. afarensis* were about 3.2 to 4ft (1 to 1.2m) tall—just under four feet in height at the maximum, with a brain only 25in³ (415cm³) in volume and ape like features; but *A. afarensis* is fully human in the way the dental arcade is fully rounded and does not have straight sides as in apes. Along with *A. afarensis*, another australopithicene termed

The skeleton of "Lucy", a member of the species A. afarensis, *throws light on one of the earliest stages of human evolution.*

Australopithecus anamensis has very recently shed light on the origins of one of the most characteristic of all human features: bipedalism. Very recent findings show that the wrist bones of *A. afarensis* and *A. anamensis* retain anatomical specializations for knuckle walking, where the wrist is bent back and weight taken on the backs of the second of the three rows of finger bones. This is critical evidence for interpreting the evolution of human gait. It suggests that the early humans went through a knuckle walking stage in which the "ancestor" was already partially terrestrial. It also shows that knuckle walking was common to both humans, gorillas and chimps rather than being a special adaptation in gorillas and chimps only.

Australopithecines such as Lucy demonstrate the possible evolutionary lineages taken in the course of the rise of modern humans. But there were other australopithecines present in Plio–Pleistocene South Africa that were very different and that must have constituted a distinct element in the terrestrial "ape" ecology of the time. *Paranthropus boisei* had a huge, flattened face with immense cheekbones, stood up to 5.5ft (1.6m) tall and probably weighed about 110lbs (50kg). *Australopithecus robustus* was similarly proportioned but did not have the "helmet" face of *P. boisei*. Neither of these animals is on the line to humans, because their jaws and teeth, which give evidence of a tough diet of grasses, show specializations such as the massive premolars and molars that preclude the possibility of their being human ancestors.

NEOGENE

THE COOLING of the Earth's climate that led to the expansion of open grasslands was the root cause of the development of a savannah-adapted land fauna, particularly artiodactyls (even-toed ungulates). In North America, a component of this fauna, which was well established by Pliocene times, was a variety of deer-like herbivores. Some of these were "true" deer or cervids, some, like *Synthetoceros* were protoceratids, a sister group to camels, and some—like the tiny *Merycodus,* only slightly bigger than the living water chevrotain—were antilocaprids, related

The savannah habitat of Pliocene North America contained an abundance and variety of ungulates.

to the bovids (cattle, sheep, and goats). The antlers of these animals were very different in shape from those of modern deer and antelope, but probably functioned in a similar way. Although similar to modern ecosystems, Neogene grassland habitats were still recognizably different and supported some bizarre creatures, such as the twin-horned *Epigaulus,* a burrowing rodent which lived in much the same way as modern marmots; its small horns are not found on any living rodent and their function remains obscure. Carnivores, too, spread out on to the open plains. In North America there were no bone-cracking hyenas, as in modern Africa, and this ecological niche was occupied by the massive-jawed "hyena-dogs" or borophagines such as *Osteoborus,* which fed on carrion.

1 *Cranioceras* (a cervid)
2 *Neohipparion* (an equid)
3 *Synthetoceras* (a proceratid)
4 *Megatylopus* (a camelid)
5 *Osteoborus* (a borophagine)
6 *Merycodus* (an antilocaprid)
7 *Epigaulus* (a rodent)
8 *Pseudaelurus* (a felid)

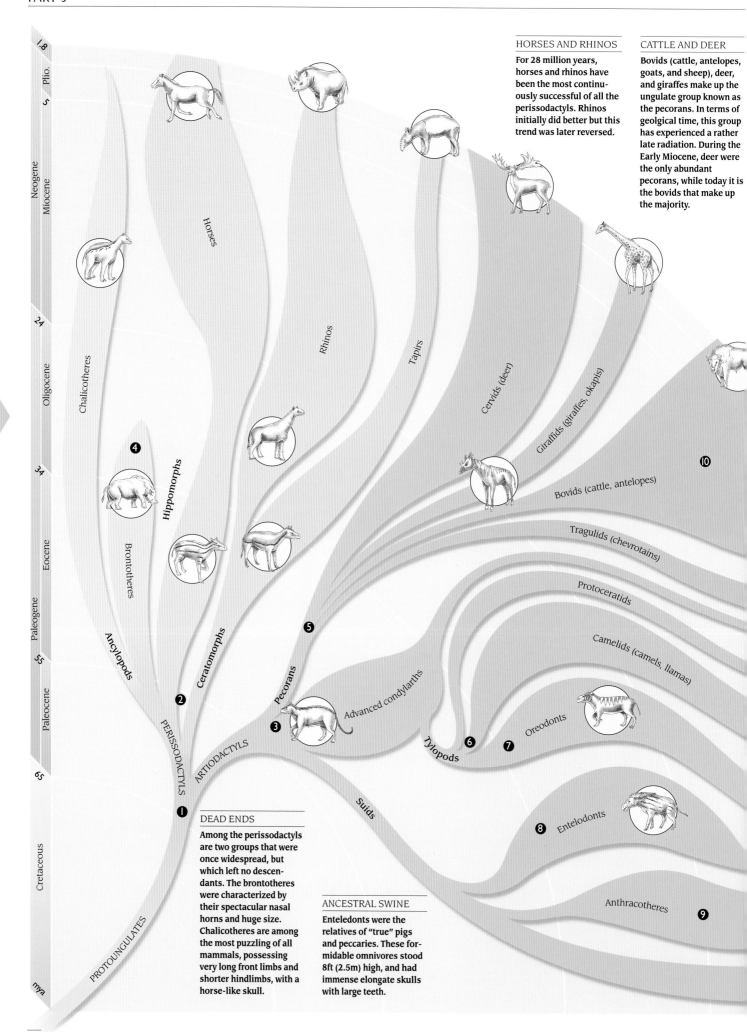

NEOGENE

1.8
Plio.
5

Neogene
Miocene

24

Oligocene

34

Paleogene

Eocene

55

Paleocene

65

Cretaceous

mya

Chalicotheres

Horses

Rhinos

Tapirs

Cervids (deer)

Giraffids (giraffes, okapis)

Bovids (cattle, antelopes)

Tragulids (chevrotains)

Protoceratids

Camelids (camels, llamas)

Oreodonts

Hippomorphs

Brontotheres

Ceratomorphs

Ancylopods

Pecorans

Advanced condylarths

Tylopods

PERISSODACTYLS

ARTIODACTYLS

Suids

Enteledonts

Anthracotheres

PROTOUNGULATES

HORSES AND RHINOS

For 28 million years, horses and rhinos have been the most continuously successful of all the perissodactyls. Rhinos initially did better but this trend was later reversed.

CATTLE AND DEER

Bovids (cattle, antelopes, goats, and sheep), deer, and giraffes make up the ungulate group known as the pecorans. In terms of geolgical time, this group has experienced a rather late radiation. During the Early Miocene, deer were the only abundant pecorans, while today it is the bovids that make up the majority.

DEAD ENDS

Among the perissodactyls are two groups that were once widespread, but which left no descendants. The brontotheres were characterized by their spectacular nasal horns and huge size. Chalicotheres are among the most puzzling of all mammals, possessing very long front limbs and shorter hindlimbs, with a horse-like skull.

ANCESTRAL SWINE

Enteledonts were the relatives of "true" pigs and peccaries. These formidable omnivores stood 8ft (2.5m) high, and had immense elongate skulls with large teeth.

THE EVOLUTION OF UNGULATES

WATER BABY

People often suppose tapirs to be relatives of the pigs, but they are far more closely related to horses. Tapirs live in the forests of South America and Asia and spend much time in the water.

Ungulates are hoofed herbivores, including cattle, pigs, tapirs, camels, rhinos, and horses—all of which have been economically important to humans for centuries. Their diversity, and the numbers of fossils and living species, testify to huge success throughout the Cenozoic. On the family tree of mammals, ungulates belong—with cetaceans (whales), aardvarks, hyraxes, sirenians (sea-cows and manatees), and elephants—to the group known as the Ungulata. There are two orders of ungulates; the perissodactyls and the artiodactyls, meaning odd-toed and even-toed respectively.

Like many other mammals, ungulates originated among one of the many groups of condylarths in the Late Cretaceous. The earliest known condylarth genus, *Protungulatum* ("first hoof-bearer"), shows a dietary shift from the pattern of other Late Cretaceous placental mammals. The cusps of its teeth are blunt and bulbous (bunodont), which improves the capacity for grinding and crushing. Early ungulates show this bunodont dentition, but later forms developed selenedont teeth with crescent-shaped cusps. Horses, in particular, evolved complex, high-crowned (hypsodont) teeth, to cope with coarse vegetation.

The artiodactyls dealt with their diet in a different way: rumination. There are several families of "true"ruminants, the cervids (deer), musk deer, tragulids, giraffids, antilocaprids (pronghorns), and bovids. Pigs, peccaries, and hippos are non-ruminants, while camels are "pseudo-ruminants." Ruminants all possess a rumen or fore-stomach from which they can regurgitate food that has been partially broken down by digestion, for another chewing inside

EVOLUTIONARY TRENDS

The number of grazers that feed on grasses and have high-crowned teeth is much greater than in the past. There has also been a decline in the numbers of really gigantic herbivores; huge animals such as brontotheres, dinocerates, entelodonts, and giant indricothere rhinos have vanished.

1 Ancestral ungulates divide into two groups
2 First adaptive radiation of perissodactyls
3 First adaptive radiation of artiodactyls
4 Gigantic brontotheres are important herbivores
5 Radiation of pecorans
6 Tylopods diversify
7 Oreodonts are the most abundant North American grazers
8 Entelodonts (stilt-legged pigs) abundant in North America
9 Amphibious anthracotheres very common worldwide
10 Huge adaptive radiation of bovids

CAMELS AND THEIR RELATIVES

Camels are one of the groups that make up the tylopods ("padded foot"). Camels evolved and diversified exclusively in North America, until the end of the Miocene, and were much more varied and abundant than today. One of the later forms, *Alticamelus*, had an enormously long neck and occupied a giraffe-style ecological niche. Swift-footed camel types remained in South America and their descendants are still there—in the form of llamas, alpacas, guanacos, and vicuñas.

Pigs, peccaries

Hippopotamuses

ODD AND EVEN

In ungulates the general mammalian foot has been modified, with the alteration of the ankle joints and the loss of one or more digits (brown). Perissodactyls have one or three toes and the animal's weight is sustained mainly through the middle toe. In artiodactyls (two or four toes) the weight is transmitted down the two middle toes. Rhino and hippo feet are splayed, for weight bearing, while in running animals the metapodial bones (blue) are elongated and fused. All ungulates walk on the tip of their toenails (hooves).

the mouth; this is known as "chewing the cud." Then the food is re-swallowed to pass through further chambers of the stomach for maximum extraction of nutrients.

The oldest artiodactyl is the Early Eocene animal *Diacodexis*, a rabbit-sized animal reminiscent of living small deer such as the muntjac, whose remains are found in North America, Europe, and Asia. Its limbs were gracile (long and thin) and were clearly specialized for running. Its body plan is echoed by all later artiodactyls.

Neogene ungulate faunas developed from Paleogene communities in a number of ways. Entelodonts, oreodonts, anthracotheres, and tapirs were all reduced in diversity and abundance, and the huge catapult-horned brontotheres had vanished. Ecological niches formerly occupied by these Paleogene animals were adopted by rhinos, deer, camels, and anthracotheres. Artiodactyls became more numerous, widespread, and diverse than perissodactyls. Today there are 79 genera of artiodactyls and only six of perissodactyls. Modern groups such as deer, cattle, and antelopes further radiated in the Late Neogene but, mostly, the main evolutionary origin of the basal forms of these groups was in the Paleogene. Origins in the Neogene were restricted to the genus level.

The standard view among scientists was formerly that the Oligocene plains and forests of North America and Asia were dominated by early horses and rhinos, while from mid-Miocene times onwards the camels, pigs, and cattle rose to prominence, but this scenario has been challenged in the last two decades. It would seem that the patterns of evolutionary radiation and extinction in both groups run more or less in parallel, rather than in opposition, and that each of these groups was evolving independently in response to environmental stimuli.

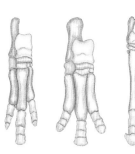

Tapir Rhinoceros Horse Hippopotamus Deer Camel

Odd-toed ungulates Even-toed ungulates

THE EVOLUTION OF PRIMATES

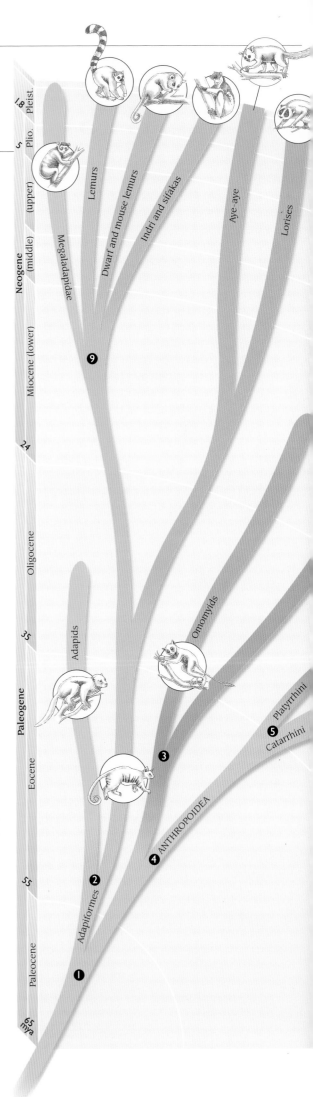

MAMMALS are characterized by their large brains and intelligence. Among them, the brain expansion of one group in particular has allowed them to occupy entirely unique ecological niches. These are the primates—including lemurs, bush-babies, monkeys, apes, and humans—which first appeared about 50 million years ago. Their special adaptations relate to agility in trees and the complex social interactions of forest life; exceptional brain-power, keen eyesight with stereoscopic vision, and extended parental care. Some are terrestrial. Most primates live in forested areas, where few fossils are preserved; as a result the fossil record of primates is not as good as that of large-bodied savannah-dwelling mammals such as ungulates, for example. Despite this, fossils do allow a glimpse of their evolutionary history.

The earliest confirmed primate is *Altiatlasius*, which is based on just ten cheek teeth from the Late Paleocene of Morocco. This tiny animal was an omomyid, one of an extinct group, mainly from the Eocene of North America and Europe, whose members looked like modern mouse lemurs or tarsiers. Most omomyids were larger than *Altiatlasius*, weighing up to 2.2lbs (1kg). The most abundant early primates were the adapiformes such as *Smilodectes*, which resembled modern lemurs and were found almost worldwide from the Eocene to the Miocene. Adapiformes, omomyids, lemurs, lorises, and tarsiers are grouped together under the heading "prosimians"—a descriptive rather than a phylogenetic term, as the group is paraphyletic, with two clearly separate lineages.

There is controversy as to whether the adapiformes or the omomyids are the basal primate group (the most anatomically primitive) in terms of primate phylogeny and which group are most closely related to the anthropoid ("human-like") suborder. Omomyids were more tarsier-like than the adapiformes; this similarity to tarsiers supports the hypothesis that tarsiers plus omomyids (tarsiiforms) are the nearest relatives of the higher primates.

EVOLUTIONARY TRENDS

The radiation of primates began in earnest at the start of the Miocene, when diversification became very rapid. Based on the anatomy of fossil primates, it is clear that ground-dwelling was a late adaptation. The origin of primates, via animals such as *Altiatlasius*, came from a small arboreal form. Terrestrial monkeys only appeared at the very end of the Miocene epoch, less than seven million years ago. Semi-bipedal terrestrial apes appeared a little after this. The development of bipedalism can be linked to the expansion of open habitats.

1. *Altiatlasius* is the oldest protosimiiform
2. The adapiformes become the most abundant primates
3. Omomyids, the most primitive group of primates, diversify
4. Anthropoids appear
5. Anthropoids split into the platyrrhines and catarrhines
6. Platyrrhines split into cebids and atelids
7. Many cercopithecoid lineages evolve
8. Initial divergence of hominoids—terrestrially adapted "apes"
9. Lemurs undergo an extensive radiation and produce a giant form—*Megaladapis*
10. Appearance of advanced hominids

PROSIMIANS

This ring-tailed lemur shows the generalized body plan of lemurs; it is an unspecialized anatomy that seems to be very similar to that of many early primates such as the adapid *Smilodectes*. Lemurs, which are all confined to Madagascar, and other prosimians, share many primate features with the anthropoids, but they also have more primitive mammalian features, such as greater reliance on their sense of smell.

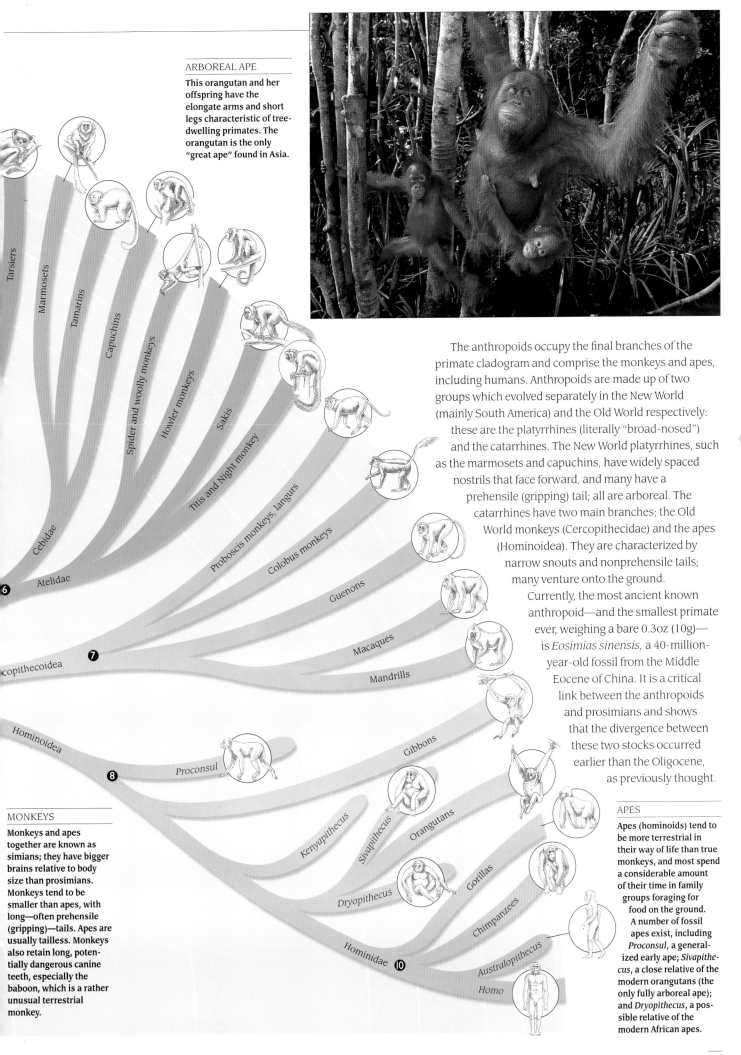

ARBOREAL APE

This orangutan and her offspring have the elongate arms and short legs characteristic of tree-dwelling primates. The orangutan is the only "great ape" found in Asia.

The anthropoids occupy the final branches of the primate cladogram and comprise the monkeys and apes, including humans. Anthropoids are made up of two groups which evolved separately in the New World (mainly South America) and the Old World respectively: these are the platyrrhines (literally "broad-nosed") and the catarrhines. The New World platyrrhines, such as the marmosets and capuchins, have widely spaced nostrils that face forward, and many have a prehensile (gripping) tail; all are arboreal. The catarrhines have two main branches; the Old World monkeys (Cercopithecidae) and the apes (Hominoidea). They are characterized by narrow snouts and nonprehensile tails; many venture onto the ground.

Currently, the most ancient known anthropoid—and the smallest primate ever, weighing a bare 0.3oz (10g)—is *Eosimias sinensis*, a 40-million-year-old fossil from the Middle Eocene of China. It is a critical link between the anthropoids and prosimians and shows that the divergence between these two stocks occurred earlier than the Oligocene, as previously thought.

MONKEYS

Monkeys and apes together are known as simians; they have bigger brains relative to body size than prosimians. Monkeys tend to be smaller than apes, with long—often prehensile (gripping)—tails. Apes are usually tailless. Monkeys also retain long, potentially dangerous canine teeth, especially the baboon, which is a rather unusual terrestrial monkey.

APES

Apes (hominoids) tend to be more terrestrial in their way of life than true monkeys, and most spend a considerable amount of their time in family groups foraging for food on the ground. A number of fossil apes exist, including *Proconsul*, a generalized early ape; *Sivapithecus*, a close relative of the modern orangutans (the only fully arboreal ape); and *Dryopithecus*, a possible relative of the modern African apes.

Tarsiers

Marmosets

Tamarins

Capuchins

Spider and woolly monkeys

Howler monkeys

Sakis

Titis and Night monkey

Proboscis monkeys, langurs

Colobus monkeys

Guenons

Macaques

Mandrills

Gibbons

Orangutans

Gorillas

Chimpanzees

Australopithecus

Homo

Cebidae

Atelidae

copithecoidea

Hominoidea

Proconsul

Kenyapithecus

Sivapithecus

Dryopithecus

Hominidae

6

7

8

10

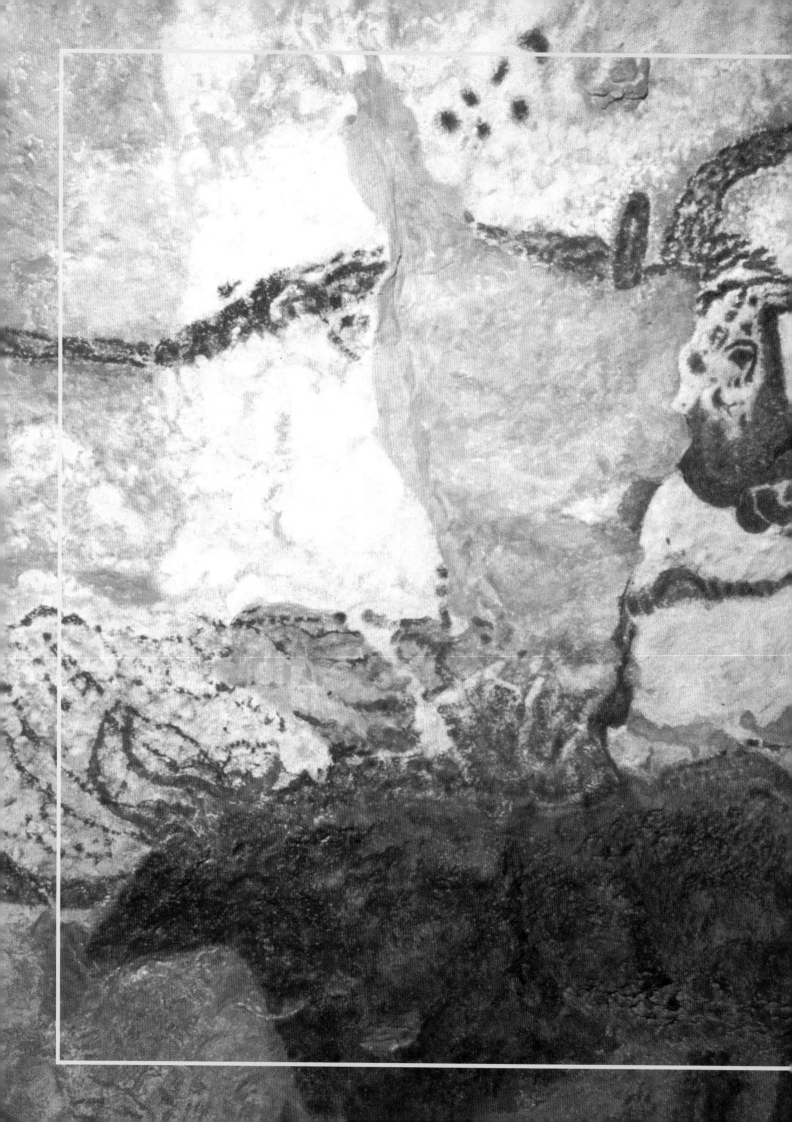

THE QUATERNARY

1.8 MILLION YEARS AGO TO THE PRESENT

THE PLEISTOCENE

THE HOLOCENE

THE QUATERNARY comprises the last 1.8 million years of Earth history. It covers the transition from paleontological science (the study of fossils) to archaeology (the much more recent time period involving the study of the remains of people and their civilisations). The adaptive radiation and colonization of the planet by the human race that took place during this period is, undoubtedly, of the greatest significance to us, its members. The impact on the Earth of human expansion is probably, at least in the short term—geologically speaking—important too. But the dominant theme of the Quaternary has been climate change.

Although the temperature of the Earth has fluctuated throughout its history, the range has been relatively small. Currently, we are experiencing one of a number of warm intervals punctuating a period of exceptional cold that has lasted for 2 million years—an ice age. Ice ages have occurred throughout geological time; there were at least four during the Pre-cambrian, one in the Ordovician and one in the Late Carboniferous and Permian. On the whole, however, the great ice ages of the past were a result of continental drift, as ancient continents moved across the poles and froze, eventually being released from the grip of ice as the geography changed again.

The present appearance of the globe, with its familiar distribution of continents and its polar ice caps is, in fact, unusual in terms of paleogeography. In the Tertiary, the separation of Antarctica caused it to become cut off from warm air and ocean currents by a circular current of cold water. The Arctic Ocean, surrounded by land-masses, also became isolated from warm currents as the Isthmus of Panama joined the continents of North and South America and produced the Gulf Stream, which carried moist air and higher precipitation to northern latitudes. The result in both hemispheres was a rapid growth in ice sheets, with those in the north advancing and retreating periodically in a way that is

The major events of the Quaternary period were widespread glaciation, particularly in the northern hemisphere, and the spread of human beings across the globe.

unique in Earth's history. Previous glacial episodes in ancient Earth history were non-repeated, isolated, b often long-lived events. Geological evidence shows t the glacial cycles that characterize the Quaternary ha never occurred before this time.

MODERN EDUCATION takes the events of glaciation as evident, but many scholars of the last century were dubious and believed that the Earth's landscape had been shaped by great c astrophes such as the biblical Flood. Similarly, until it was realized that ani extinctions had occurred not just onc but repeatedly, fossil remains were attributed to animals that had been drowned. Invoking divine action to explain natural phenomena was a normal idea in nineteenth century society, given the the philosophical attitudes and limited scientific tech-niques of the time. A glacial theory wa only accepted in the 1830s, when the Swiss-American naturalist Louis Agas provided robust interpretations of geological feature such as the embankments of debris termed moraines *roches moutonnées* ("sheep rocks") with their nubby texture, and hanging valleys whose open ends drop i larger valleys—all testament, not to a deluge, but to t power of glaciers to permanently sculpt the landscap

Quaternary glaciations also affected the oceans. A the ice sheets waxed and waned, there was a resultin dramatic rise and fall in sea level—as much as 300ft (100m). The cyclic deepening of global oceans was reflected in the formation of deep-water black shales and siliceous sediments resulting from an increase in cold-water radiolarians—silica-shelled microscopic planktonic organisms. Successive changes in sea lev are visible in stepped coral reefs on the shores of the Caribbean islands. Reefs and beaches formed at different times were exposed when sea levels fell, or covered when sea levels rose, with corals that grew in shallow water now found deep below the surface.

IN GEOLOGICAL TERMS, the world of the Quaternary was much as it is today, with all its geographical features in nearly identical positions. But when the continental glaciers were at their furthest extent, with ice extending in Europe as far as northern Italy, and in North America as far as New York, the fall in sea levels caused land-bridges to appear. The Bering land-bridge that connected Siberia and Alaska and allowed the passage of animals between Asia and North America is no longer present, but fossils from Quaternary sediments document the past existence of two-way migrations between Asia and North America. Similarly, the European ice sheets allowed the migration of animals to the British Isles, where the fauna during the Quaternary was similar to that of mainland Europe.

During the periodic retreat of the ice sheets, or interglacials, these connections were lost and the biotas developed differently. However, the temperate conditions of the interglacials (such as we are currently enjoying) temporarily allowed warm-adapted animals to move north. The presence of fossil lions, hyenas, and hippopotamuses in Britain shows how warm these interglacial periods were. The geology of the British Isles records repeated transitions from Arctic steppe-tundra through boreal birch and conifer forest to temperate broad-leafed forest and back again. It is likely that, in less than 10,000 years, we shall be under several hundred feet of ice cover once again.

Land-bridges also made possible the rapid spread of modern humans from their origin in East Africa. By the beginning of the Holocene, the time of permanent settlement and the development of agriculture, all other hominid species were extinct and the world population was 4–5 million; today it is about 10 billion. People have become a powerful force on the Earth. We have gained a great deal of knowledge about its past and its workings but, so far, we have not used that knowledge particularly wisely. Our actions in just the last 150 years have brought the extinction rates of animals to an all-time high and are now having an impact on the global climate. We should consider ourselves as stewards of the planet if we are not to prove the undoing of huge numbers of its creatures—ourselves included.

THE
PLEISTOCENE
1.8 – 0.01
MILLION YEARS AGO

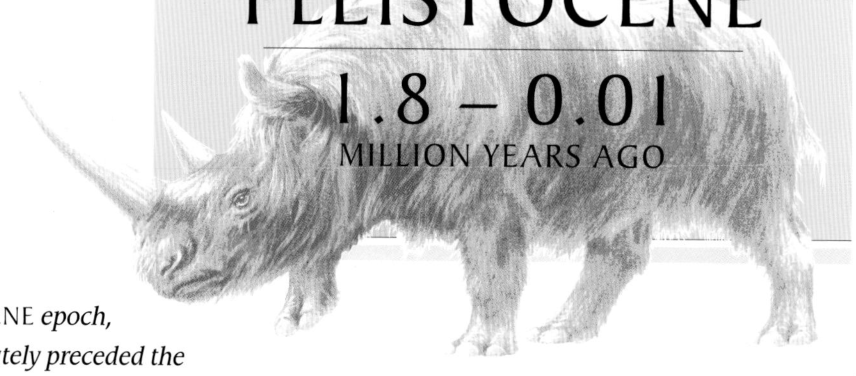

THE PLEISTOCENE *epoch, which immediately preceded the current Holocene, lasted from approximately 1.8 million years ago until about 10,000 years ago. Among the scientific disciplines it spans the transition from paleontology (the study of very ancient life forms as fossils) to archaeology (the study of the remains of humans and civilization) as primitive humans evolved into more advanced beings. These humans developed complex social structures that gave rise to civilized societies, which have left tangible archaeological remains.*

To paleontologists and geologists, the most characteristic feature of the Pleistocene was the repeated cycle of glaciation and deglaciation as huge ice sheets advanced, covering a third of the northern continents, and then retreated. Other epochs had had episodes of glaciation, but what was unique in the Pleistocene was the rapid reversal between glacial and interglacial conditions. As a direct result, there were many repeated extinctions of plants and animals, both terrestrial and marine. These glaciations also caused major redistributions of life.

ON TODAY'S Earth, approximately 14 percent of the total land area is covered by ice sheets or underlain by ice-cemented rock, and about 4 percent of the entire ocean surface is shrouded by a thin layer of ice that fluctuates seasonally. During the past 2 million years, as much as 25 percent of the land and up to 6 percent of the oceanic surface area have been covered or underlain by ice. This global ice mass is referred to as the cryosphere and is composed of sea ice, permafrost, and glaciers. The Pleistocene glacial events represent an entirely different set of climatically controlled processes from those that had existed for much of Earth's history. Because the Pleistocene glacial cycles have occurred only in the past 2 million years or so, their effects have been superimposed on much older rocks. Glacial geological features are found in all areas that have been subjected to the effects of glaciers, regardless of the nature of the underlying strata. As the ice sheets formed and then melted, alternately locking up and releasing immense volumes of glacial water, changes in terrestrial water supply caused the formation of many distinctive glacial landscapes. There were repeated episodes of various land and lake redistributions, one legacy of which we see today—for example, in the Great Lakes of North America. Many of these landscape features are geologically young.

> *Ice covered 25 percent of the land and 6 percent of the oceans—three times more ice than today.*

KEYWORDS

BERING
LAND BRIDGE

GULF STREAM

HOMO SAPIENS

INTERGLACIAL

MITOCHONDRIAL EVE

MORAINE

OUT OF AFRICA
HYPOTHESIS

PRECESSION

TUNDRA

NEOGENE	1.8 mya	1.7	1.6	1.5	1.4	1.3	1.2	PLEISTOCENE
Series								EARLY/LOWER
European stages								CALABRIAN
North American (mammalian) stages								IRVINGTONIAN
Glacial periods (Europe)			DONAU			Donau/Günz interglacial		GUNZ
Glacial periods (N. America)								
Geological events				Ice sheets cover about 30% of continents in the northern hemisphere				
		Continuing subduction of Pacific plates beneath N. American plate						
Sea level								Very shallow, fluctuating
Archaeological periods	LOWER PALEOLITHIC (Oldowan)							Oldest use of fire
	First *Homo ergaster*							
Animal life	• First *Smilodon* saber-tooth cats		• First mammoth					

Sir Charles Lyell, the great British geologist, defined and named the Pleistocene in his historical four-volume book *Principles of Geology*, which was first published in 1833. The Pleistocene was formerly named the Newer Pliocene, a name that stood until 1837. Lyell based his definition of the Pliocene epoch on its marine fossil faunas. Pliocene strata contained fewer species that are alive today than the more recent Pleistocene epoch. The Swiss–American paleontologist Louis Agassiz, on the other hand, was the first person to realize that the very characteristic landforms of Europe were the result of glaciers in the region, and he did not define any particular epochs as Lyell did. Agassiz is remembered as a paleontologist and geomorphologist, Lyell as a general geologist and theorist.

Many nineteenth-century naturalists were skeptical about glaciation, and usually called on biblical accounts of floods and other divinely-directed phenomena to explain unusual geological features such as "erratic" boulders that lay far from their origins. It was only when a substantial amount of incontrovertible evidence was compiled by the geologists of the time that natural explanations for glacial geology superseded religious interpretations. It was in the 1830s that Agassiz, in the face of much opposition, insisted that glaciers had been responsible for forming much of the landscape of the modern world, from the steep valleys of his native Switzerland to the Great Lakes of North America.

The Plio–Pleistocene glaciation of the northern hemisphere also provided an explanation for the disjunct distribution of some species, such as the magnolia tree, which had puzzled naturalists. Moving south through Europe ahead of the ice sheets, this plant had its progress blocked by the glaciers that had developed in the Alps and Pyrenees and disappeared, leaving populations in the Americas and Asia. With the realization that glaciations could redistribute populations, the significance of these geographical curiosities was finally appreciated.

D RAMATIC environmental changes on land were associated with the latest episode of the Pleistocene ice ages, which involved the growth and decay of very large ice sheets in the mid-latitudes of the northern hemisphere. These episodes included ice sheets covering North America and the Canadian Rocky Mountains, northwest Europe and the British Isles, the Arctic Islands off the north shores of Canada, and much of the northern regions of Siberia. During the last glacial period, about 18,000 years ago, the huge ice sheets in northwest Europe and in North America eventually reached their maximum extents. Following an initial period of very slow decay, they began to decay very rapidly at about 14,000 years ago. This stage stopped sometime between 11,000 and 10,000 years ago, but then continued. The ice had more or less disappeared in Europe by about 8,500 years ago and in North America by about 6,500 years ago.

Along with the growth and extension of large ice sheets in the northern hemisphere, many of the earth's principal environmental zones, or biomes, were displaced toward the equator. Immediately south of the ice sheets in the northern

Dates of the glacial cycles are approximate, because movement of the ice varied across the globe.

See Also

THE PALEOGENE: *Psychrosphere, Grasslands, Carnivores*
THE NEOGENE: *Himalayas, Panamanian land-bridge, Hominids*
THE HOLOCENE: *Caribbean, Andes, East African rift valley, Glacial relicts*

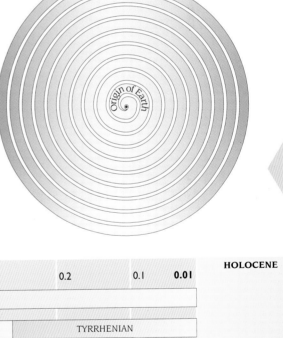

PLEISTOCENE

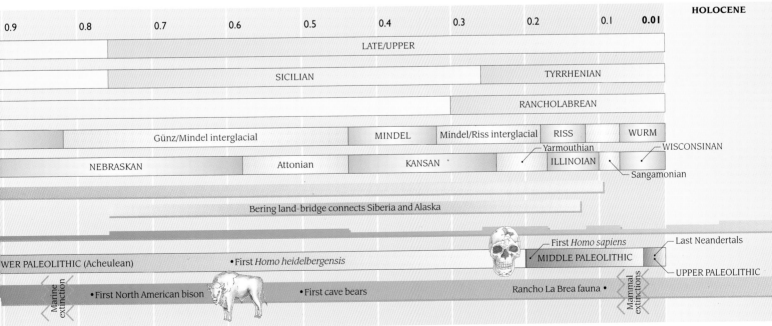

71

GLACIATION AND ISOSTATIC ADJUSTMENT

When ice sheets expand during glacial events, they cause both the lowering of sea levels through a decrease in water volume as ice mass increases on the continents, and isostatic depression of the land beneath them. Isostasy is the mechanism by which the Earth's crust stays in gravitational equilibrium as it floats on the denser, more plastic mantle beneath. So, in the same way that icebergs are supported by seawater—the ice below the surface weighing less than the water it displaces—the large volume of less dense rock in continental crust gives it buoyancy, with the thicker mountain regions being balanced by deep roots. As mountains erode, the root rises in compensation. Similarly, when ice sheets retreat there is gradual uplift of the depressed crust as the load is removed. This can be considerable: as much as 1100ft (330m) over the last 10,000 years in the Hudson Bay area of Canada, for example. However, such uplift is a delayed response, and a large amount will remain to be completed after the total withdrawal of the ice sheet. As the ice melts during warm interglacial periods, sea levels rise and flood into the isostatically depressed areas; the Baltic Sea is one result of this. The relationship between the rising and falling of sea levels and continents is complex, but it results in a characteristic geological feature known as raised shorelines. A "staircase" of these raised beaches can be found in some places high above current sea levels. Raised shorelines are particularly evident in northern Canada, Scandinavia, and Spitsbergen, Norway, all of which were heavily loaded by Pleistocene ice sheets and are still rebounding even now.

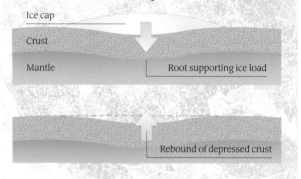

Ice cap

Crust

Mantle

Root supporting ice load

Rebound of depressed crust

hemisphere lay the tundra. This sparsely vegetated zone, where winter temperatures may be as low as –134°F (–57°C), extended as far south as northern France in Europe and at least 100mi (150km) south of the ice sheet margin in central North America. Tundra is found today, particularly in the Siberian steppes and the Canadian Arctic; vegetation there rarely grows more than a few inches high and is low in nutrients. Several inches below the soil surface is permafrost, a layer of ice which remains permanently frozen, even during the summer. Ice-age tundra was probably an even harsher place, and the permafrost extended many tens or hundreds of feet below ground level. Here, unable to penetrate, surface water cut deep channels as it ran off.

As the immense ice sheets expanded southwards from the Arctic regions, the latitudinal variations in climate were compressed towards the equator as ice covered the north. Climates thus varied more over shorter distance in a north–south direction than before. Biological zones were located closer together and there was less distance between distinct faunas. Climate zones shifted south, bringing rain to what had been (and have become once again) arid regions, such as eastern Africa.

It was here that the first humans appeared approximately 300,000 years ago, an event that may prove to be one of the most consequential of all in the history of the planet. At the time, however, humans were merely a timid new entry onto the evolutionary scene, which was dominated by large, ferocious, swift-moving carnivores. "By all the biological laws, these ill-armed, gangling beasts [humans] should have perished," in the words of the biologist Loren Eiseley; but they flourished instead.

A MAP of the Pleistocene world is essentially the same as that of today, although some differences are evident. For example, the Bering land-bridge that stood between Siberia and Alaska was in existence and—importantly—allowed passage between Asia and North America. Many of the fossils from Pleistocene deposits record a transfer of animal types between the two continents. This land-bridge existed on and off from about 75,000 to 11,000 years ago and would have been covered with tundra vegetation. Enormous herds of grazing animals such as musk ox and reindeer moved across the bridge into Alaska and Canada. Although the continental ice sheets were widespread, at times there was an ice-free corridor to the west, down which the first humans to reach America traveled in pursuit of the great herds.

Ice sheets covered most of the northern hemisphere, but beneath them the distribution of the land masses was much the same as it is today.

The advancing ice sheets diverted the north Atlantic Gulf Stream to the south, forcing it towards Spain, and the Mediterranean "sluice" between the straits of Gibraltar, making northern Europe still cooler. In places farther afield, the drop in sea levels as a result of water being locked into the ice sheets, had geographical effects as some islands were reconnected to their nearest continents. The area of the Caribbean islands, for example, was increased as the drop in sea levels exposed more of their previously submerged offshore land areas. At the height of the last glaciation the sea levels around Barbados were as much as 400ft (120m) lower than today.

WEST COAST
TECTONICS

In western North America, regional uplift produced the Rocky Mountains. Along the coast, subduction and volcanism continued along the San Andreas Fault.

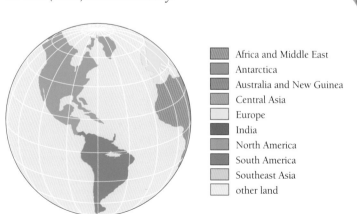

Africa and Middle East
Antarctica
Australia and New Guinea
Central Asia
Europe
India
North America
South America
Southeast Asia
other land

TWO TYPES OF ICE

The northern hemisphere was covered in both pack (sea) ice and continental ice. Sea ice forms in a relatively short period, but continental ice sheets grow over hundreds of years during episodes of intense cold.

Bering
Landbridge

rn Corridor

Laurentide Ice
Sheet

Scandinavian
Ice Sheet

EUROPE

Mediterranean
Sea

NORTH
AMERICA

AFRICA

a Madre Occidental

n Trench

Caribbean Sea

Panamanian
Landbridge

ATLANTIC OCEAN

Amazon
Lake
System

Peru–Chile Trench

SOUTH
AMERICA

DIVERTED WATERS

By the time the Panamanian land-bridge had been in existence for three million years, its redirection of warm Atlantic currents carried moist air to northern latitudes, increasing precipitation and freshwater runoff into the ocean. This resulted in the rapid growth of continental ice sheets.

CONTINENTAL
GLACIER

The only present-day
continental glaciers are
over Antarctica and
Greenland. Glacial ice
(left) flows from the ice
sheet into the sea at
Jakobshavn in
west Greenland.

In the east of the globe, mountain glaciers covered the newly uplifted Himalayas and extensive meltwater lakes lay in the vast areas of Eurasian tundra. Although the most dramatic glaciation was in the north, the Antarctic ice sheet expanded to about double its present area, with a great increase in the amount of sea ice. The circumpolar current that had formed in the late Paleogene dragged the pack ice eastwards. Small ice sheets grew in the Andes and the mountains of Australia and New Zealand. At this time of low sea levels Australia was joined to New Guinea; the East Indies were also continuous land, with extensive river systems. Their modern island shapes appeared as mountain ranges.

APART from the periodic waxing and waning of the huge ice sheets, which had occurred in other periods (notably the Ordovician and the Permo–Carboniferous),

> *The modern ice age is a complex event, with pulses of glacial expansion (maxima) separated by partial retreats (minima).*

most physical parameters of the Earth's atmospheric, hydrospheric, and biotic systems had long been established in patterns that continue today. With the Pleistocene, a number of sudden temperature falls occurred. It was clear, by the end of the nineteenth century, that the concept of an "ice age," which had replaced that of "the deluge," was not a simple event, but that there had been more than one glacial episode. Eventually a chronology of four major glaciations and three interglacials was widely accepted. Evidence for these episodes comes from geology, geochemistry (oxygen isotopes in Greenland ice cores), and fossils, and even tree-ring dating techniques. Since the 1950s, radiometric dating has enabled fairly precise dates to be obtained.

The last glacial maximum (LGM)—episode of widespread ice cover—took place in the Pleistocene epoch about 20,000 years ago, but in subsequent millennia the earth warmed and the ice sheets of the northern hemisphere retreated northward into the ancient Arctic circle. Our planet is currently in the middle of an interglacial period. Such periods last only a short time but have been shown to have extremely fast rates of biotic and physical change—in fact, faster than ever before.

Climatic instability typifies the closing stages of a glacial period and the opening of an interglacial, and shows how fast climates can change in certain circumstances. This instability was first noted in cyclic lake sediments in Denmark. The sediments dating from the transition between the glacial and current interglacial exhibited some rather unusual features: tundra was replaced as vegetation stabilized the shores of the ancient lakes and productivity increased in the water column. But this process, reflecting the increasing warmth of the climate, was evidently interrupted. The clays show a return to cold climates as heavy clays became evident again in the sediment cores; these features were found in cores of the same age throughout the entire European continent.

The cold episode was severe enough to cause the regrowth of the northern glaciers, which left their mark on Scandinavia and Scotland. This cold spell was called the Younger Dryas event because the fossil leaves of *Dryas octopetala*—mountain aven, an Arctic–alpine plant—were found in abundance in the clay layers. Radiocarbon dates show that the Younger Dryas took place between about 11,000 and 10,000 years ago. At a number of sites in continental Europe, a shorter period of cold again took place about 12,000 to 11,800 years ago; it is called the Older Dryas.

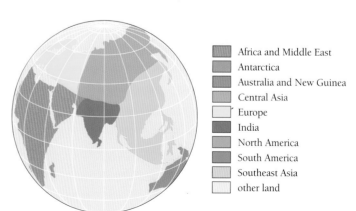

Africa and Middle East
Antarctica
Australia and New Guinea
Central Asia
Europe
India
North America
South America
Southeast Asia
other land

Taymyr Ice Sheet

ASIA

JAPAN

INDIA

EAST INDIES

NEW GUINEA

INDIAN OCEAN

PLEISTOCENE

PLEISTOCENE TUNDRA

In Asia, tundra steppes lay at the southern margins of the ice sheet. These immense flat expanses received meltwaters that subsequently filled geographical depressions and formed extensive lakes.

AUSTRALASIA

Australia was connected to New Zealand and New Guinea during the early stages of the Pleistocene epoch. The northward drift of the Australian plate caused the uplift of seamounts that eventually formed the Indonesian island chain. At this time, though, a large amount of land was above sea level.

ORBITAL VARIATION

The Earth's orbit varies in its eccentricity (departure from circularity), tilt (inclination of the axis of rotation to the orbital plane), and precession (wobble, as in a child's spinning top).

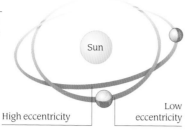

Sun

High eccentricity   Low eccentricity

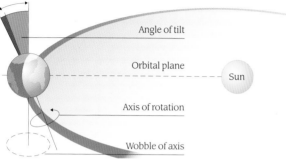

Angle of tilt

Orbital plane   Sun

Axis of rotation

Wobble of axis

THE EARTH'S atmosphere, oceans, biosphere, and ice sheets are part of one huge global system in which these components are very strongly coupled. A change in any one part will lead to changes in the others. Each of the different components of the global climatic system operates on an entirely different timescale. The atmosphere adjusts to changes very quickly, more so than the other components—this occurs over a period of just weeks, whereas the oceans and the biosphere adjust more slowly, over periods of hundreds to thousands of years, and ice sheets are slowest of all to respond. Changes to ice sheets tend to occur over tens of thousands to hundreds of thousands of years.

> *Ice ages are slow-moving phenomena, taking many thousands of years to respond to global changes.*

The major expansion of global ice cover that occurred during the late Ordovican and the Permo–Carboniferous was connected to the changing position of the super-continent of Gondwana in relation to the South Pole. Glacial conditions migrated from North Africa in the late Ordovician, to South Africa during the Carboniferous, and to Australia during the Permian. Because climatic changes brought about by plate movements occur extremely gradually, on a scale of millions of years, plate tectonics theory cannot be used to explain the repeated cycles of glaciation that are are uniquely characteristic of the Pleistocene epoch.

This phenomenon puzzled early geologists. In 1876 the British geologist James Croll suggested that long-term changes in the amount of solar radiation reaching the Earth were controlled by rhythmic variations in its orbit around the Sun, and that these were responsible for the periodic climate changes that European geologists had only begun to discover. However, it was not until 1941, when the Yugoslavian astronomer Milutin Milankovitch calculated the magnitude and frequency of the changes in solar radiation received by the Earth as a result of orbital changes, that there was a mechanism to account for Croll's proposal.

Milankovitch identified three orbital processes that would control these changes: the tilt of the Earth's axis, the eccentricity of the Earth's orbit, and the precession of the equinoxes. The Earth's axis is not at right angles to the plane of the Earth's orbit about the Sun, but inclined at an angle of about 23.5°. This angle varies from between 24.5° and 21.5° in cycles of about 40,000 years; when it is greatest, there is the largest difference in seasonal heating at any latitude.

The Earth's orbit is not circular, and sometimes it is more strongly elliptical than other times. The length of this eccentricity cycle is about 100,000 years. The Earth

ICE FLOE

Pack ice is usually only a few meters thick. Ocean movements can cause it to break up, with ice at the edge of the pack floating free as floes (above, off the coast of Spitsbergen). Ice floes drift under the influence of winds, unlike icebergs, which move with ocean currents.

MEASURING THE ICE AGES

Foraminiferans (forams) are microscopic single-celled animals that form a globular, calcareous skeleton around their cell wall. They form part of the ocean plankton, and, upon death, sink to the sea floor to accumulate along with other, benthic (bottom dwelling), types. Over millions of years they are preserved in sea bed lime-stones that often seem to contain nothing but forams. Oxygen is present in sea water as the stable isotopes oxygen-16 and oxygen-18, and these are incorporated into the skeletons of forams in the same proportion in which they occur in the surrounding sea water. The relative proportions of these two oxygen isotopes in the fossil forams found in deep-sea cores fluctuates widely, with the ratio shifting towards heavier values during glacial maxima.

Initially, scientists thought that these fluctuations reflected variations in water temperature; however, it was found that similar fluctuations were exhibited not only by planktonic forams but also the deep sea benthic forams. This is the region in the oceans of the freezing-water layer that has existed since the formation of the psychrosphere 35 million years ago. It is now known that temperature has only a minor effect on the oxygen–isotope ratio in the foram skeleton, and that this fluctuation rather reflects the oxygen isotope ratio of the oceanic waters in which the foram lived; which varies greatly with the size of continental ice sheets.

During glacial episodes, more of the lighter oxygen-16 falls as snow. This accumulates in the glaciers, leaving a higher proportion of the heavier isotope in the ocean. Recognition of the fact that this high proportion of oxygen-18 in foram skeletons corresponds with an increased volume of glacial ice was a critical moment in establishing the use of forams (such as the planktonic *Globorotalia* species, right) as paleoclimatic indicators. Correlation with periods of magnetic reversal, as shown by geomagnetic analysis of the rock cores, calibrates the chronology of climatic cycles that foram isotope analysis displays. In this way, the waxing and waning of the Pleistocene ice sheets can be precisely dated.

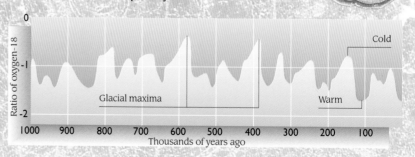

Ratio of oxygen-18

Cold

Warm

Glacial maxima

0   -1   -2

1000   900   800   700   600   500   400   300   200   100

Thousands of years ago

Feedback mechanisms in climatic systems have major effects. Growth of ice sheets increases the albedo or reflectivity of the Earth's surface, with a net cooling effect larger than the consequence of Milankovitch cooling alone. Cooled waters absorb more carbon dioxide than warmer waters and decrease the greenhouse effect, further contributing to global cooling. Other effects include the partially isolated Arctic Ocean in the northern hemisphere, which began to spill its cold waters into the north Atlantic as Greenland rifted off North America. In the southern hemisphere, the circumpolar Antarctic current had been supplying cold global waters in the form of the psychrosphere for more than 40 million years. All of these effects have combined to produce the extraordinary glacial cycles of the past 2 million years, although the exact mechanism by which this occurs has yet to be fully explained.

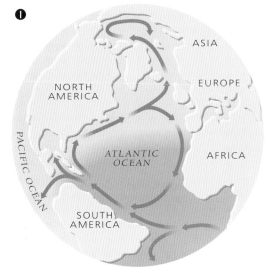

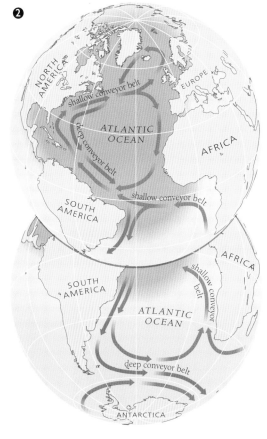

also wobbles slightly on its axis—a phenomenon known as precession— due to the gravitational effects of the Moon and Sun on the Earth's equatorial bulge, which changes the timing of the solstices. These are relative to the position that the Earth occupies in its elliptical path around the Sun. About 11,000 years ago, the Earth's nearest point to the Sun occurred when the northern hemisphere was tilted towards the Sun (northern hemisphere summer), rather than during the northern hemisphere's winter as is the case today. The length of the precession cycle is about 23,000 years.

Analysis of the very long-term variations in the Earth's climate show that the pattern of change between 800,000 years ago and today is composed of three dominant rhythms: one with a period of 100,000 years, one of 40,000 years, and one of 20,000 years. From this it can be concluded that variations in the Earth's orbit are the main determinants of long-term patterns of climate.

However, Milankovitch cycles cannot be reconciled with some geological data. The temperature change that takes place between the coldest parts of glacials and the warmest parts of interglacials is about 4–5°C, but the difference in the intensity of solar radiation reaching the Earth due to orbital variations is not enough to change the global average temperature by more than 0.4–0.5°. There are also changes in the climatic rhythms through time: before 800,000 years ago, the 40,000 year rhythm was the governing rhythm and after this time the 100,000 rhythm was dominant. Finally, there is no fundamental change in the "orbital pacemaker" which might explain the progressive intensification of the ice ages over the past 2 to 3 million years. As a result, there is not yet a completely satisfactory explanation for why the Pleistocene glaciation cycle began.

ATLANTIC CIRCULATION

One possible cause of Quaternary glaciation was the changing circulation pattern of the Atlantic Ocean. During the early Pliocene (1), before the Isthmus of Panama formed, warm Atlantic waters flowing north mixed freely with Pacific waters. This would have made the Atlantic waters less saline, and therefore more buoyant, than today. They may then have flowed into the Arctic Ocean, warming the polar regions, before cooling made the current dense enough to sink. The creation of the isthmus cut off access to the Pacific, and the drying effect of the trade winds blowing off the Sahara increased the salinity of the Atlantic. Today (2) these denser waters sink north of Iceland, forming a loop known as the oceanic conveyor belt (lower maps), isolating and cooling the whole Arctic region as it brings cold water down from the North Pole to the South in deep-water currents.

➡ shallow ocean current
➡ deep ocean current
▨ high-salinity water

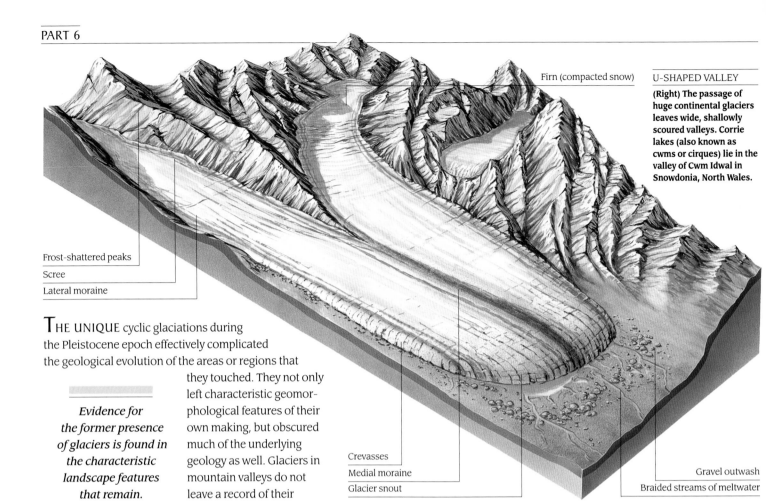

Firn (compacted snow)

Frost-shattered peaks
Scree
Lateral moraine

Crevasses
Medial moraine
Glacier snout

Gravel outwash
Braided streams of meltwater

U-SHAPED VALLEY

(Right) The passage of huge continental glaciers leaves wide, shallowly scoured valleys. Corrie lakes (also known as cwms or cirques) lie in the valley of Cwm Idwal in Snowdonia, North Wales.

THE UNIQUE cyclic glaciations during the Pleistocene epoch effectively complicated the geological evolution of the areas or regions that

Evidence for the former presence of glaciers is found in the characteristic landscape features that remain.

they touched. They not only left characteristic geomorphological features of their own making, but obscured much of the underlying geology as well. Glaciers in mountain valleys do not leave a record of their presence in the rock strata because these highland regions are areas of erosion rather than of deposition. In contrast, continental glaciers do leave their mark. An example of this is seen in modern Greenland and Antarctica, which are almost entirely covered by vast ice sheets.

The Greenland and Antarctic glaciers are thickest at their centers. Gravity causes them to flow outwards from these high points as the typically riverlike glaciers that are more familiar in less remote places, such as the Alps. These glaciers, which flow in geographical channels, are called valley glaciers when they are very long relative to their width and may prove to be confluent with other valley glaciers. Cirque glaciers (or "corries" in Gaelic and "cwms" in Welsh) are relatively small bodies of ice that occupy hollows high on the sides of mountains. They

GLACIAL EROSION

(Above) Ice is usually considered a solid, but behaves as a slow-moving liquid, influenced by gravity and slope. In mountain glaciers there is gradual downward flow of ice. The erosive force of moving glaciers is evident from the landscape that they leave behind: U-shaped troughs, hanging valleys from tributary glaciers left suspended, sharp-edged arêtes, isolated horns—produced as cirques grow at the glacier head—and debris such as till and moraine.

GLACIAL ERRATICS

(Left) Glaciers entrain, carry, and deposit rocks of different geological provinces from distant regions; often from as far as 300 miles away (500km). These "erratics" appear as randomly scattered, slightly striated boulders that are usually of recognizably different lithologies to those of the surrounding valley strata. Although the source of most erratics is unknown, the origin of some can be determined. By studying them, it is possible for geologists to trace the path of an ice flow.

occur in places where the temperature is only just low enough to maintain a permanent snow field (firn), and where there is insufficient snow and ice to support a valley glacier.

The geological evidence of past glaciations is represented by a suite of features that result from the peculiarities of glaciers. Glaciers move rocks, erode rocks, and move sediments in their direction of flow. Rocks entrained at the base of a glacier erode the substrate still further to give highly characteristic scratches on rock surfaces that are called glacial striations. When glaciers flow over rock "knobs" they tend to pluck blocks from the lee side of the knob due to pressure release at this location of the rock. These plucked blocks are entrained within the bottom region of the glacier where they then act as grinding tools by cutting striations into the bedrock. The combination of lee-side plucking and up-glacier side-smoothing of bedrock knobs produces the widespread rockforms of smoothed and striated hummocks known as *roches moutonnées*.

Uplifted areas provide high-velocity streams that cut deeply into rock. In complex glacier systems, where relatively minor tributary glaciers feed into more rapidly flowing trunk glaciers, the trunk (main) glacier tends to erode down into the bedrock much more rapidly than the tributary (side-branch) glacier. As a result of this, the trunk valley is deeper than the tributary valley, which now forms a hanging valley in which its open end is high up

on the side wall of the trunk valley. V-shaped valleys are eroded by valley glaciers into U-shaped valleys because the glacier erosion occurs across the whole cross-section of the valley instead of being concentrated by rivers only on the center line of the valley.

MORAINE is transported debris that had originally been entrained by a glacier and has now been dumped following its retreat. Moraines mark the present or former positions of ice margins. They come in various types, such as end, or terminal moraines, where material carried by the ice is deposited at the terminus of the glacier, while recessional moraines are formed as the ice retreats, each representing a temporary halt. Lateral moraines occur as ice erodes the sides of a valley; when it melts, the accumulation of rubble is left next to the valley walls and medial moraines are created where the lateral moraines of two valley glaciers join. If the ice advances across a moraine the sediments can become contorted and folded; this feature is called a push moraine.

Cape Cod on the northeast coast of North America is an example of an unusually large medial moraine.

Moraines are scattered over large areas of both North America and Eurasia. Some of these moraines are of spectacular sizes; one extends out into the north Atlantic as Cape Cod. Terminal moraines often flank depressions in the country rocks of a landscape, and were made as the glaciers retreated. Some of these huge depressions became the Great Lakes of North America. The Hudson Bay of Canada is not bounded by a terminal moraine; it is an arm of the ocean that spread into a region where the crust had been depressed by the thickest part of the continental glaciers in North America, and has not yet rebounded back to its higher original level.

During the erosion of a subglacial rock bed, the process of crushing and abrasion during glacial transport produces debris with a wide range of sediment grain sizes, from silt to large boulders; this may be deposited subglacially as a sediment termed till. Unlike moving water and wind, ice cannot sort the sediment it carries, so till is characterized by being composed of unsorted mixtures of many particle sizes; in addition many pieces are striated and polished by the action of the glacier. Till that has become lithified due to burial, compaction, and chemical alteration is termed tillite. Both tills and tillites are extremely characteristic of the past influences of glaciers.

Glaciers tend to block normal river drainage paths; during the recession of the Pleistocene ice sheets, ice-dammed lakes came into a temporary existence. However, permanent lakes also receive meltwaters and glacially transported sediments. Glaciers are often

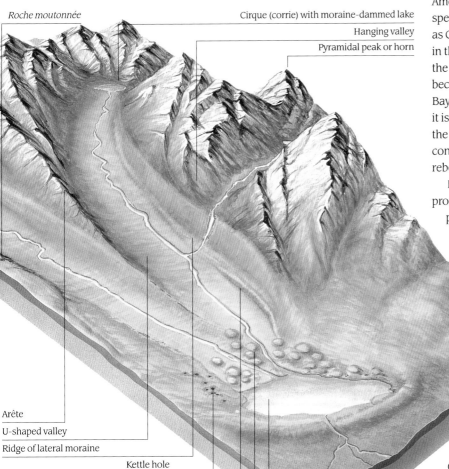

Roche moutonnée

Cirque (corrie) with moraine-dammed lake

Hanging valley

Pyramidal peak or horn

Arête

U-shaped valley

Ridge of lateral moraine

Kettle hole

Esker

Drumlins

Moraine-dammed lake

Outwash plain

influenced by seasonal climatic changes; as winter approaches, meltwaters cease as they are locked up as ice and the input of water into lakes diminishes, but upon warming in the spring, meltwaters flow freely again and fill up the lakes. This annual cycle leads to the accumulation of annual layers, called varves, in sedimentary rocks. Furthermore, icebergs containing debris "calve" (divide) into many glacial lakes, and any debris entrained within them is released as they melt. These released rocks, called dropstones, drop into the finer lake-bottom sediments. They are a sure indicator of previously glaciated terrains and may be easily identified in the field.

POSTGLACIAL LAKES

Ice sheets over large areas of the Canadian Shield disturbed drainage patterns. Today, postglacial lakes (right) fill hollows that were eroded by ice or blocked by moraine left by retreating ice sheets. (Below) Other lakes, such as Lake Bonneville, developed in basins south of the glaciers; these pluvial lakes resulted from the increased runoff and rain.

EVOLUTION OF THE GREAT LAKES

(Right) The Great Lakes are the remnants of huge ice lobes that extended southwards from the margin of the Pleistocene ice sheets. The ice sheets formed depressions that were subsequently filled with meltwaters during their retreat about 14,000 years ago (1). By 10,000 years ago (2), all the modern Great Lakes had begun to fill except Lake Superior, whose basin still lay beneath the ice. The Great Lakes are just a few of an enormous number of lakes formed by continental glaciation in the shield region of North America. More lakes were created in this area by glacial processes than by all other geological processes combined. However, the size of the Great Lakes greatly surpasses that of all the others.

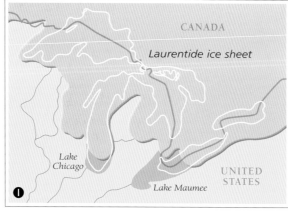

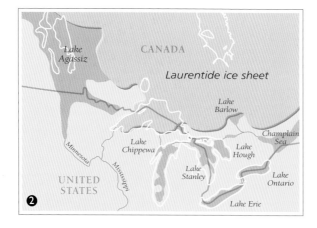

SCIENTIFIC investigations of Pleistocene and Holocene geology in the region of the Great Lakes, and examinations of their possible origins, began in the mid-nineteenth century. The Swiss–American paleontologist Louis Agassiz of Harvard University went to Lake Superior in 1850 and described some aspects of its geology in great detail. It was Agassiz who was the instigator of the ideas regarding the mechanisms of glaciation and its effects on the land: these effects had left a characteristic geomorphology, and Agassiz was the first to recognize the classic signs. His researches were followed in later years by Lawson (1893) and F.B. Taylor (1895 and 1897), both of whom studied the glacial geology of the Great Lakes area. Agassiz gave his name to the largest of the ancient proglacial lakes (lakes formed directly adjacent to the snout of glaciers), Lake Agassiz, which covered a vast area of south-central Canada, with an arm extending south into North Dakota.

The Great Lakes of North America (and similar lakes elsewhere in the world) were formed as a result of Pleistocene glaciation.

The age of the Great Lakes was unknown in the last century, and despite immense advances in geological science, is still largely unknown. This is because they were successively excavated by the ice sheets, and each excavation obliterated any previous evidence that might be used to date the system. However, evidence from rocks surrounding the Great Lakes shows that these inland bodies of water were not in existence prior to the Pleistocene epoch. This means that they are not more than

PLEISTOCENE

Pleistocene occurred in North America only, and this was about 800,000 years ago, at a time when Europe was somewhat less glaciated. The development of changing topographies in and around the Great Lakes system caused the formation of different lakes at different times throughout the Pleistocene epoch. Examples of these now-vanished ancient lakes are Lake Milwaukee, Lake Leverett, Lake Maumee, and Lake Saginaw.

The geomorphology of the terrain that surrounds each lake is different and gives each one its own particular shape. Such individual peculiarities of the lake geomorphology also extended to glacial lakes in Europe; here much smaller basins were formed which now include the Saalin basin in the Netherlands and the Elsterian basin in Germany. However, the sheer size of the American continental plate meant that there was a much greater potential to form extensive glacial lake systems than in the more circumscribed area of Europe. This area also features a structure that has become famous: Niagara Falls. The dramatic waterfall at Niagara, which lies between Lake Erie and Lake Ontario, came about when the retreating ice of the last glacial uncovered an escarpment that was formed by southerly tilted, mechanically resistant strata (dolomite, a hard limestone). Water from the Niagara River flowed over the edge of this escarpment— which is supported underneath by the solid Lockport Limestone series—and undermined the weaker shales below, causing a southward, upstream erosional retreat of the falls. The Pleistocene glaciations that covered North America have given this area a number of magnificent landscape features, such as the Great Lakes and Niagara Falls, yet these natural wonders are still quite young in geological terms.

about 1.7 to 2 million years old. The present floors of the Great Lakes were previously lowlands. The glaciers that were extending southward from their northern site of origin flowed slowly into these lowlands, scouring them deeply and isostatically depressing the crust (by up to 1100ft/330m in the Hudson Bay area). As the glaciers withdrew, over a period of 10,000 years or so, during the interglacial periods, freezing meltwaters collected in these depressions and began to form the present Great Lakes system.

During the glaciation and meltwater events, huge blocks of ice enclosed in glacial clastics collected to the south of the lake basins and gradually began to melt. As a result of this mosaic pattern of melting ice, a characteristic hummocky topography was formed in the regions of what is now mid-northern Michigan. This unusual landscape is composed of irregular hills and small lakes known as kettles. At a later stage in the Pleistocene epoch, sands that had been deposited by the melting waters along the southern shores of lake Michigan were later carried eastwards by North American weather systems. These sands were deposited as dunes around the southern bend of the lake, where some can still be seen today. Because the resulting Great Lakes are so big, their shorelines show tilt when measured, and even rates of isostatic uplift can be determined. The lakes also controlled ice movements by providing more channels of lower relief than the surrounding countryside into which the glaciers preferentially flowed due to the effects of gravity. The geological history of the Great Lakes after the Wisconsinian glacial stage is a complex one of changing meltwater outlets and the presence or absence of ice barriers. The Wisconsinian stage glaciations of the

THE LAKE MISSOULA FLOOD

(Right) The channelled scablands of Washington State, USA, are a complex of interlaced deep channels called coulees. They were formed when a large ice lobe advanced southward and blocked the Clark Fork River—a major tributary of the huge Columbia River. (1) Ice dams trapped waters, which banked up to form Lake Missoula in Western Montana. As the glacier retreated, the ice dam broke, releasing a vast and unparalleled flood; the water swept across the Columbia plateau, stripping away soil and cutting deep channels (2).

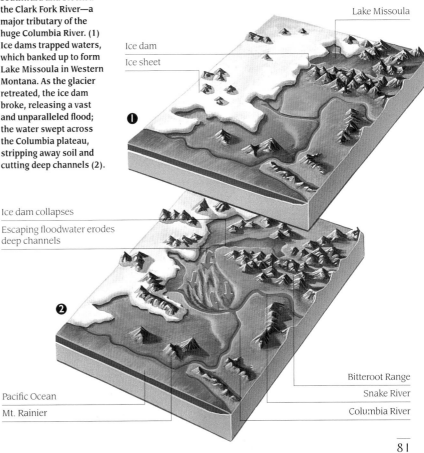

Lake Missoula

Ice dam

Ice sheet

❶

Ice dam collapses

Escaping floodwater erodes deep channels

❷

Pacific Ocean

Mt. Rainier

Bitteroot Range

Snake River

Columbia River

THE GLACIAL cycles of the Pleistocene epoch brought about great differences in the geographical distributions of mammals during the cold and warm phases as the ice sheets waxed and waned. This is particularly well documented for small mammals such as rodents and mustelid carnivores (weasels, ferrets, and stoats). In the latter part of the Pleistocene, many mammals that had formerly occurred together became separated due to the increased patchiness and disjunction of their habitats. These tundra faunas are classed as "disharmonious " faunas because they contain animals that no longer co-exist. Tundra steppes probably extended to a much greater area than today and supported a fauna of large, medium, and small animals, all of which were adapted to the harsh, cold environments, or could migrate away from the winter tundra with ease. The expansion of the Pleistocene ice sheets had a marked effect on animal life in the northern hemisphere, and one of these effects was the formation of an extensive tundra biota, the legacy of which still remains with us today.

The tundra was inhabited by a wide range of cold-adapted animals.

The tundra of today's planet is found enclosing the Arctic Circle, north of the tree-line, and in smaller areas in the southern hemisphere on the sub-Antarctic islands. Alpine tundra also occurs above the tree-line on high mountains, including those in the tropics. For tundra vegetation there is an extremely short growing season and only cold-tolerant plants can survive; typical tundra plants are mosses, lichens, sedges, and dwarf trees. The main large animals that live there are reindeer (caribou) and musk ox. Small herbivores include snowshoe hares, voles, and lemmings. In addition, many birds migrate to the tundra from the south in the summer in order to feed on the huge warm-season insect population.

Feeding on these are the carnivores, which in modern tundra ecosystems are represented by Arctic fox, wolves, falcons, hawks, and owls. Fossils of animals that derive from what appears to be Pleistocene tundra steppe environments include a number of animals that still inhabit this ecosystem, such as reindeer, elk, hares, wolves, voles, and ermine. The latter two mammals lived under the thin seasonal coverings of snow rather than attempt to burrow into the iron-hard soil to any great extent.

The tundra of Pleistocene times was more widely populated than today's, and by animals that are no longer seen anywhere on Earth. Some of the most significant of these were not native. In response to interglacial–glacial oscillations, mammals either migrated or expanded their ecological tolerances to climatic and/or vegetational changes. Associations of some mammals with certain types of environments and plants suggest that many have altered their preferences or tolerances. For example, the hamster is now found on Asian steppe but has been recovered as fossils from sediments deposited in forested environments during interglacial periods. Modern wild horses are typically associated with open environments, but fossil wild horses are also found in forested habitats, as well as grasslands. Canids had their evolutionary origin in North America and crossed the Bering land-bridge into Asia where they continued their success as predators of open habitats. Those animals that immigrated into North America across the land-bridge were cold-adapted, such as mountain sheep, musk oxen, moose, lions—and humans. Oxen of different types were especially prevalent, but mammoth elephants were less common.

MAMMAL RANGES

(Right) The traditionally migratory herbivores such as elk, woolly mammoth, and woolly rhinoceros had much greater biogeographical distributions than most contemporary Pleistocene carnivores, possibly as a result of greater adaptability. Saber cats were dominant in North America, but Europe was inhabited by dirk-toothed and scimitar-toothed cats.

MAMMOTH BONES

(Above) The widespread incidence of mammoth remains, such as these in South Dakota, has provided good information about their appearance. The Siberian form was smaller than the American, with males being about 10ft (3m) high at the shoulder.

PLEISTOCENE

Woolly mammoths are usually imagined scratching a living in snowy wastes devoid of vegetation. This is misleading, for such an environment could probably not support herds of these animals, with their enormous food requirements. It seems likely that woolly mammoths inhabited the boreal (forested) regions outside the tundra, where food would have been available at least part of the year; and also the warmer and more equable margins south of the steppes, where a richer vegetation was more readily available. Very short mosses and lichens could not have been grazed by tall-standing mammoths, and their trunks could probably not have gained a grip on small low-lying vegetation. Oxen, on the other hand, with their downward-pointing necks, grazing dentition, flexible lips, and broad muzzles, were particularly well adapted to graze efficiently on the extremely short tundra vegetation. This factor partially accounts for their abundance in these harsh environments. The same functional consideration of ecology holds true for the reindeer and elk. These large-bodied animals provided considerable bounty for proficient carnivores such as wolves that were capable of bringing them down in the hunt.

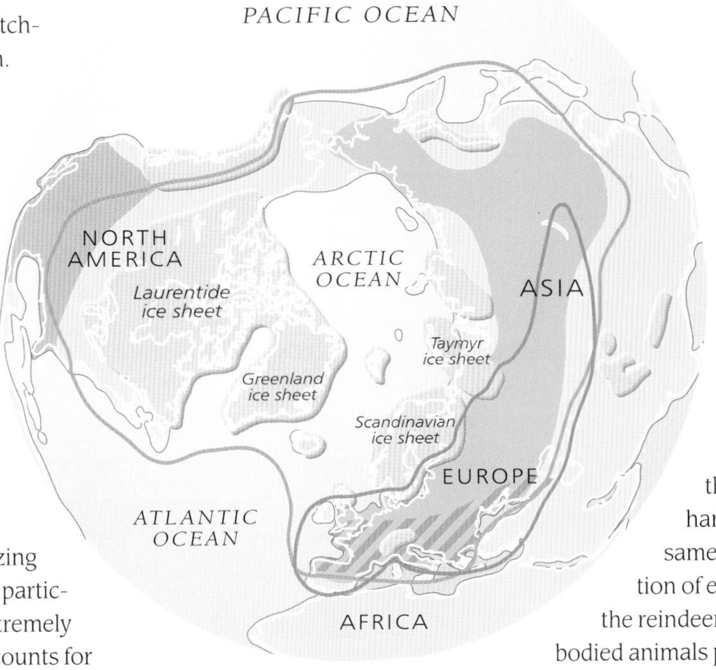

SEVERAL hundred miles south of the tundra margin, in the region which is now California, lay a radically different biome. Despite the northern glaciation, California was still relatively warm, and it was home to its own distinctive fauna. Some of these are familiar today, while others are mysterious and often spectacular extinct representatives of modern animal groups. There is a considerable amount of information about this rich ecosystem due to an important geological feature on the west coast of North America: the San Andreas Fault. This fault is one feature of a major system that extends in a broadly north–south

The San Andreas fault is the visible boundary between the North American and the Pacific plates.

PACIFIC OCEAN

NORTH AMERICA
Laurentide ice sheet
ARCTIC OCEAN
ASIA
Taymyr ice sheet
Greenland ice sheet
Scandinavian ice sheet
EUROPE
ATLANTIC OCEAN
AFRICA

Upper Pleistocene mammals
- giant deer
- woolly mammoth
- cave bear
- woolly rhinoceros
- saber-toothed cat

— coastline, 12,000 years ago
ice cap, 12,000 years ago

WOOLLY COATS

(Below) Large-bodied animals lose heat more slowly than smaller creatures, so that large body size is generally advantageous in cold climates. Nonetheless, animals the size of mammoths required extra insulation against the extremes of the Pleistocene ice ages. Smaller animals may burrow or hibernate, but huge ones cannot do this. Continuous exposure to such cold conditions resulted in the super-thick, hairy coats of mammoth, bison, and rhinos.

1 *Coelodonta antiquitatis* (woolly rhinoceros)
2 *Megaloceros giganteus* (giant deer)
3 *Mammuthus primigenius* (woolly mammoth)

❸

PLEISTOCENE

trend along the western side of the whole of the Americas. This huge line, several thousand miles long, is the demarcation zone of collision between the constituent plates of the Pacific lithosphere and that of the American continent. From the north these Pacific plates are: the Kula plate, the Farallon plate, the Nazca plate, the Phoenix plate, and, to the extreme southern tip of South America, the Antarctic plate. As the spreading ridge of the East Pacific Rise pushes it east, the subduction of the Nazca plate beneath the South American plate is largely responsible for the formation of the Andes Mountains.

In the California area of the eastern Pacific, the North American lithospheric plate is heading south while the Farallon plate that forms the northeastern margin of the Pacific plate goes north. These plate margins exist as the San Andreas transform fault and scrape past each other, over a stretch of land several hundred miles long. This is known as a strike–slip fault, or one in which the two opposing blocks of lithosphere are passing sideways to each other, not over or under either block. Occasionally, over time, the contacts between the rock surfaces of these two lithospheric plates lock and grab, and as a result of the continued plate movements they flex and bend before snapping back into position. The release of this massive stress is felt as earthquakes. Measurements show that the plates are moving at a velocity of about 2in (5cm) per year; the past average was 0.4in (1cm) per year. According to these figures, Los Angeles should be parallel to San Francisco in 25 million years or so.

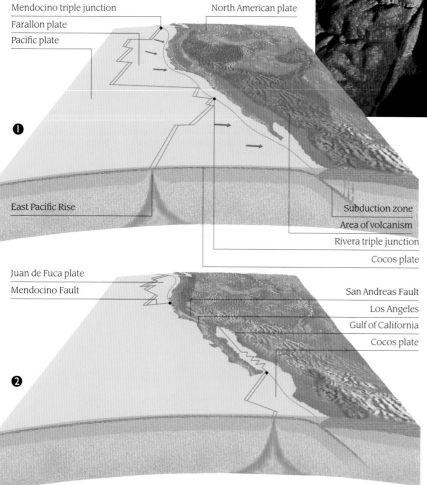

❶
Mendocino triple junction
Farallon plate
Pacific plate
North American plate
East Pacific Rise
Subduction zone
Area of volcanism
Rivera triple junction
Cocos plate

❷
Juan de Fuca plate
Mendocino Fault
San Andreas Fault
Los Angeles
Gulf of California
Cocos plate

THE MAJOR fault zone of the San Andreas Fault is restricted to a very narrow belt, sometimes described as a mobile belt, extending south into the Gulf of California where there are a number of smaller, stepwise transform faults in the seabed. Strike–slip faults such as the San Andreas also form shallow basins as offset portions of the two lithospheric plates stretch apart, causing subsidence between them. These terrestrial basins receive sedimentary input from the surrounding areas and act as centers of deposition; consequently they have a high potential to preserve the remains of the animals that lived in those low-lying regions. The San Andreas Fault extends through the lithosphere. Many of its smaller microfaults, extending outwards into the surrounding rocks of both the partly undersea Farallon plate and the terrestrial North

> *Fractures in the Earth's crust allowed seepage of tar to reach the surface. This collected into pools in which many animals met their deaths.*

SAN ANDREAS FAULT

(Left) In some places, the exposure of the San Andreas transform fault is particularly evident. The relative motions of the plates can be determined by careful mapping and measurement of the separate sides of the fault. It is at times when the two sides of the fault grip, lock, and part explosively that earthquakes occur.

EVOLUTION OF THE SAN ANDREAS FAULT

(Below left) About 25 million years ago the East Pacific Rise began to be subducted under the North American plate (1). As the rise was consumed, the Farallon plate was reduced to two remnants and the San Andreas Fault developed, lengthening as the Rivera and Mendocino triple points moved north and south. Between 4 and 3 million years ago (2), the fault migrated inland and a segment of Mexico—including Baja California—became fragmented and joined to the Pacific plate. The San Andreas and Mendocino faults are both transform faults. The Mendocino connects the Juan de Fuca spreading ridge to the subduction zone beneath the North American plate, and the San Andreas connects the Juan de Fuca Ridge and a second divergent boundary in the Gulf of California.

OIL DEPOSITS

(Below) At Rancho La Brea an earlier series of faulted Tertiary deposits is unconformably overlain by Pleistocene deposits. Impermeable layers in the Tertiary rocks have allowed oil to collect; it escapes to the surface via numerous fractures.

Asphalt pits
Unconformity
Pleistocene strata
Oil sands
Tertiary sediments
Fracture

American plate, intersect underground oil reservoirs. These reservoirs are the result of the accumulation of decaying organic material. An abundant animal community has left its trace in the form of decaying matter, which typically collects in low-lying areas such as estuaries and deltas before becoming entombed under layers of deposition.

Under the ground at depth, chemical changes due to pressure, temperature, and entrapment change the decaying organic matter into oil. The semi-liquid organic matter is then released at the surface due to the rupture of the underground oil reserves. These are mostly located in permeable beds, some of which are at shallow depths in geologically recent Pliocene deposits. Oil oozes to the surface through these faults in the sediments, gradually becoming more viscous on contact with the air, as the result of oxidation and the loss of its volatile constituents, and becomes tar, or asphalt. It pools at the surface, where (because it is water-repellent) it is covered by rainwater. These water-covered tar deposits look superficially like waterholes, but any animal that comes to drink at them runs a serious risk of becoming trapped. Fur and feathers are easily clogged by the tar, and animals may stray too far out into the waterhole in order to escape predators or to take a drink.

Due to the continuous shifting of the ground in the region of the San Andreas Fault, a Pleistocene tar pit now lies exposed in the middle of urban Los Angeles. The tar pit at Rancho La Brea is perhaps best known for its spectacularly preserved mammals such as giant ground sloths, mammoths, bison, large-headed "dire" wolves, and truly gigantic condor-like vultures that are known by the appropriate term of teratorns or "terror birds." But the species most often associated with Rancho La Brea is the saber-toothed cat *Smilodon*. It is usually represented as lion-sized, with unusually long teeth, but recent analysis has shown some unexpected features of

this highly successful predator. It was up to 50 percent heavier than a fully-grown lion, and the bones of its front limbs were twisted in response to a massively expanded front limb musculature. Shorter hind limbs made it incapable of running down its prey in the same manner as a modern lion. Instead, this animal relied on brute strength to bring down its prey—mammoths and ground sloths—which were far bigger than any modern lion could tackle. The great strength of the forequarters of *Smilodon* and its remarkable teeth functioned together: it grasped its huge slow-moving prey in a powerful immobilizing embrace to minimize any movement that could break the saber teeth. These were plunged into the prey and used in conjunction with the lower jaws to lever a huge chunk of flesh out of the prey in what has been called the saber-toothed shear-bite. The victim rapidly bled to death—almost immediately if the bite were on the neck.

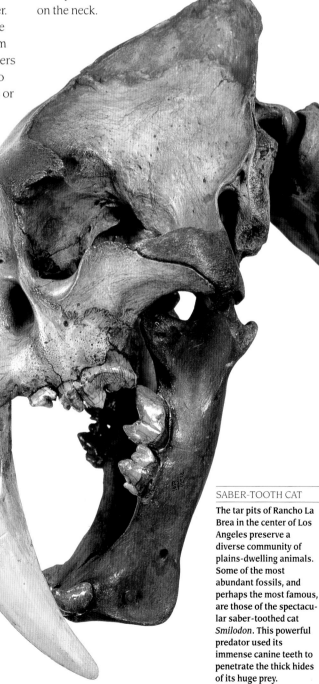

PLEISTOCENE

SABER-TOOTH CAT

The tar pits of Rancho La Brea in the center of Los Angeles preserve a diverse community of plains-dwelling animals. Some of the most abundant fossils, and perhaps the most famous, are those of the spectacular saber-toothed cat *Smilodon*. This powerful predator used its immense canine teeth to penetrate the thick hides of its huge prey.

ISLAND LIFE: BIG AND LITTLE

The Panamanian Isthmus turned South America from an immense island into a conjoined continent with North America, bringing previously isolated animals into contact with invaders The isolation of animals on islands has a major impact on their evolution; some biological features of this are characteristic and can be identified in the fossil record. The evolution of miniaturized animals is one such characteristic and is a major demonstration of how evolution can work. Geographical isolation allows a gene pool to become different from the "normal stock" of animals from which the island types descended while still in contact with the original continental population. However, the key to understanding island dwarfing is to consider the effects of a confined area in which many requirements for life are reduced to critical levels: a drastically reduced foraging and breeding space, diminished supply of nutrients and fresh water, and limited terrain in which to escape predators. Generally, island biological effects are most marked on terrestrial animals such as mammals and reptiles; birds are less prone to geographical isolation by a sea barrier.

In terms of evolution, isolation works wonders: it leads to dwarfing of big animals and gigantism in small ones. Herbivores may become small because of limited food resources, but small animals such as rodents become large only when stranded on islands that have an absence of predators. The Pleistocene epoch was the time of the giant lemur *Megaladapis* on the island of Madagascar, a bear-sized animal that fed on fruits and leaves and evolved as the result of a lack of really large predators. A living example of gigantism is the Komodo Dragon, the largest species of lizard. An absence of large mammalian predators on the Komodo Islands of Indonesia has resulted in its becoming the top carnivore there.

The effect of island gigantism is complemented by the effect of island dwarfism, such as in the dwarf elephant, whose fossils are found on the island of Malta. Elephants stranded during the Pleistocene, facing few predators, developed dwarf strains, some no bigger than a St. Bernard dog. Other examples are dwarf mammoths on Wrangel Island off the coast of Siberia, and the more widely known Shetland pony.

ANOTHER direct effect of the subduction zone between the Nazca and South American plates is the Isthmus of Panama. This land connection between North and South America—which makes up Central America—is essentially a mountain chain that is more or less confluent with the Andes mountains in South America: the mountain-building episode that uplifted the Andes also threw up the Panamanian isthmus. However, the connection between North and South America was not a complete one, but rather a series of long islands and promontories stretching between these two continents.

The formation of the Gulf Stream has had major implications for the climate, ecosystems, and habitats of western Europe.

The formation of the Panamanian land-bridge only became complete during the Pliocene, about 3 million years ago, with a major impact on terrestrial vertebrate biotas of both North and South America. There were equally momentous effects that had impact on distant regions of the planet. Up until this time, the huge oceanic currents of the Atlantic ocean had flowed from east to west (due to the rotation of the Earth and its atmospheric directions) and passed more or less unhindered from the Atlantic Ocean into the Pacific Ocean by way of the open seaway between the islands of the Central American mountain chain. But when the Panamanian Isthmus completely formed it blocked the passage of Atlantic waters into the Pacific and redirected these warm equatorial waters northeastwards along the edge of North

WATER BOUNDARY

(Above) This image of the Atlantic Ocean from space shows the boundary between the warm, fast flowing water of the Gulf Stream (lower half) and the cooler, calmer coastal waters of the eastern USA. The Gulf Stream has been crucial in the development of northern hemisphere ecosystems. It has not only kept Europe much warmer than its high latitude would otherwise indicate, but also prevented a southward extension of the sea ice during the Pleistocene ice ages. Therefore, the North Atlantic ice sheet never spread into southern Europe.

America. This diversion is the basis of the warm ocean current known as the Gulf Stream that bathes western Europe and the British Isles.

The importance of the circulation of the Earth's ocean currents coupled with those of the atmosphere explains the remarkable rapidity with which some of the climatic changes of the Pleistocene epoch have come about, and the initiation of the Gulf Stream is just one example of this. The warmth of the north Atlantic Ocean arises from warm, saline waters traveling north at intermediate depths of about 800m, eventually surfacing near Iceland. They then cool and sink (cold water being denser and

THE IRISH DEER

(Right) The so-called "Irish Elk" was, in fact, a deer; but with antlers spanning as much as 10ft (3m), it was truly a giant. The skull is unusually wide in order to accommodate the stresses imposed by such huge antlers, as are the neck bones. The antlers alone must have required a considerable energy intake to maintain their health and growth.

PLEISTOCENE

ISLAND CLIMATES

Hippopotamus fossils from British interglacial Pleistocene deposits (c.120,000 years ago) testify to the heating effects of the Gulf Stream on this region, but by 20,000 years ago, glacial ice had spread over Scotland, northern England, and most of Wales and Ireland, with a corresponding change in the fauna. The fossil remains of cold-adapted mammals such as bison, reindeer, and elk have been found in some abundance—evidence of a much colder climate than today. Lower sea levels allowed passage of these mammals from mainland Europe, and occasional land-bridges allowed some British mammals to cross into Ireland, including, by the end of the period, the Irish giant deer *Megaloceros* (below).

IRELAND

GREAT BRITAIN

—— northern limit of hippo (last interglacial)

—— southern limit of ice sheet (last glacial advance)

➡ possible land bridges

Quaternary mammalian sites
- limestone areas with cave deposits
- river terrace deposits
- bogs and fens
- marine crags

heavier than warm water). Before the Isthmus of Panama was completed the water would have been less saline, having mixed with water from the Pacific, and would have reached the Arctic Ocean before sinking. When the isthmus closed, the drying effects of the trade winds led to greater evaporation and higher salinity, which caused the water to sink sooner, thus isolating and cooling the Arctic. This cooling may have initiated the ice age.

During the Pleistocene glaciations, the formation of immense ice sheets in the north had extremely significant effects on the Gulf Stream which at that time had been in existence for less than one-and-a-half million years; initially ice floes occupying much of the north Atlantic pushed the Gulf Stream towards western Europe and the Iberian peninsula. However, during the glacial maxima, very strong latitudinal temperature gradients set up by the formation of extensive continental and oceanic ice sheets strengthened the trade winds of the northern hemisphere and pushed the warm equatorial currents that formed the basis of the Gulf Stream back into the southern hemisphere; these currents were also blocked from reaching Europe by the southward extension of the fully-formed ice sheets. Fossil evidence gained from marine invertebrates such as forams and radiolarians, from Atlantic sediments, suggest that the "conveyor belt" of currents that initiated the formation of the Gulf Stream were essentially switched off during major glacial episodes, further reducing heat transfer to the north and contributing to the formation of ice sheets.

During those times when the Gulf Stream was at its most active, Europe and the British Isles were kept much warmer than their northerly position would otherwise allow—an effect that is still very much with us today. It also, at times, had a major impact on animal life in the British Isles, the climate of these small islands and their mammalian fauna changing rapidly and dramatically under the competing influences of the cold polar front and the warm Atlantic waters.

Faunal diversity was greater on the European continent than in Britain which, in turn, was greater than that of Ireland. British temperate/interglacial faunas— prior to the last interglacial—are very similar to continental faunas. Until the early mid-Pleistocene there was also contact between Ireland and mainland Britain via land-bridges, as indicated by the distribution of large carnivores such as hyenas and bears. The earliest indications of British isolation are in the last interglacial (120,000 years ago), in which animals found in Europe—pine voles, extinct rhino, horses, and humans— are absent in Britain. British Last Cold Stage (LCS) faunas were similar to those of Europe, suggesting unimpeded migration. But the Irish LCS faunas are impoverished (no woolly rhinos or humans) and imply no connection with mainland Britain. Continued isolation, and thus faunal impoverishment, is indicated by the lack in Ireland today of such animals as voles, frogs, and snakes, which are present on the mainland.

BECAUSE of the physical nature of the environments in which early humans lived, and the low preservation potential of these regions for fossils, the evidence for human evolution is patchy. The early stages, especially, are poorly known, and theories for the origins and subsequent dispersal of anatomically-modern humans are among the most contentious of all scientific

About 1.9 million years ago in Kenya, a new, taller, upright hominid appeared.

arguments of any discipline. However, to begin with, most paleoanthropologists agree that there are two separate lines of hominid evolution, the australopithecines and *Homo*, which took place in east and southern Africa between approximately 5 and 2 million years ago. All these hominid species followed the same basic design: they were essentially small bipedal apes with small brains, large cheek-teeth for processing tough plant foods, and slightly curved hand and foot bones—suggesting a partly arboreal lifestyle.

Then, about 1.9 million years ago, in the latest Pliocene, a new hominid species arose, the fossils of which showed significant anatomical advances over those of the fossil hominid *Homo habilis* (the oldest

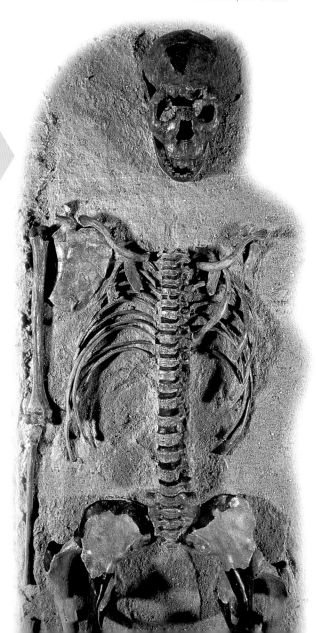

PLEISTOCENE

MITOCHONDRIAL EVE

The mitochondrion is one of the semi-autonomous organelles found within the eukaryotic cell and contains its own DNA. Mitochondria DNA is passed on only through the maternal line (red). Lineages that just produce males (blue) disappear, so that, in time, it is possible for a population of diverse MtDNA types to be replaced by a descendant population of only one type, through a process of random variation.

fossils of which are close to 2 million years old). The most perfectly preserved specimen of this new hominid species was collected in 1984 by Richard Leakey near Lake Turkana in Kenya. The specimen—known as "Turkana boy" and given the species name *Homo ergaster* ("workman"), though originally called *Homo erectus* because of its upright stance—was that of an adolescent male hominid. He stood about 5.3ft (1.6m) tall and had a brain volume of around $51in^3$ (830cc). The skull was clearly more primitive than that of *H. sapiens* (modern humans) in its large eyebrow ridges, heavy lower jaw, wide nasal opening, rounded facial region, and lack of a bony chin. Nevertheless, the skeleton appears to be more or less modern, and it walked upright on its two legs. *H. ergaster* sites show that this hominid manufactured advanced tools and weapons, ate meat, and that it foraged and hunted in groups in a cooperative manner. Similar fossils have been found in North Africa, Asia, Indonesia, and Europe, mostly dating from about 1.25 million years ago, and it is these later specimens that are assigned to *H. erectus*. In 1995 a specimen of *H. erectus* was found in China. It has also been dated at 1.9 million years, suggesting that the species has a longer evolutionary history than previously thought and that *H. erectus* emigrated from Africa closer to 2 million years ago—twice as long as was once believed.

Homo ergaster/erectus improved on the technology of its predecessor, *H. habilis*. Most of the tools used by the latter are simple and rough, consisting of no more than rounded pebbles, typically with only a single cutting edge; they are Oldowan tools, named for the site at Olduvai Gorge in East Africa. A number of *Homo erectus* sites in Europe have produced characteristic sharp-edged tools—known as Acheulean, after Saint-Acheul in France—and associated human remains that date from about 780,000 to 530,000 years ago. The presence of the tools may indicate that during the Middle Pleistocene of Africa and Europe there was a unique radiation of humans that were more derived than *Homo erectus*, but ancestral to subspecies of *Homo sapiens* of Europe.

OUT OF AFRICA

(Right) Modern humans, *Homo sapiens*, arose in Africa c.150,000 years ago and dispersed throughout the Old World and into Australia c.50,000–35,000 years ago. As they did so, they probably caused the disappearance archaic humans such as *H. erectus*. Some 18,000 years ago the last glacial was at its height, with maximum spread of ice sheets and low sea levels. These conditions were an added catalyst for the development of a diversity of specialized hunting techniques, tool and weapon development, and survival strategies.

"UPRIGHT MAN"

(Left) *Homo erectus* ("upright man") was the first geographically widespread human being. Its skeleton shows anatomical advances such as a larger cranial volume, suggesting a larger brain. This skeleton, of its ancestor *H. ergaster*, which is one of the most complete early hominids ever found, stands about 5.3ft (1.6m) high and is that of a male, as shown by the pelvic structure. He appears to have been about 12 years old when he died.

IN 1987 and 1991, studies of the molecular DNA of *Homo sapiens* made some fundamental claims: that all modern humans are closely related, that is, they share a common ancestor; that the ancestor originated in Africa; that this origin occurred no more than 200,000 years ago; and that archaic species made no contribution to the modern gene pool. The studies relied on that portion of our genetic material known as mitochondrial DNA (mtDNA), which is found in the mitochondria of cells. MtDNA evolves more rapidly than nuclear DNA; it can be used as a molecular clock to show the differences between populations due to accumulated genetic mutations that have taken place since a past point of divergence. The longer two populations have been separate, the greater the amount of genetic difference. MtDNA is inherited only from one's mother, so the hypothetical ancestor became known as "Mitochondrial Eve" and the hypothesis the "Out of Africa" hypothesis.

Studies of mitchondrial DNA show that the first modern humans originated from a common African ancestor, then spread to all parts of the globe in a new wave of migration.

The scientists examined gene diversity among, initially, 147 people from different parts of the world and then a further 189 people, including 121 Africans from six sub-Saharan regions; they found that there was only about 0.4 percent variation among the individuals (indicating a recent origin), with a greater diversity among the Africans (reflecting their longer period of evolution). By comparison to the genetic variation among our closest relatives, the apes, this is very small, suggesting a very recent divergence of modern human populations. Both studies proposed that modern humans arose in Africa around 200,000 years ago, with a common female ancestor from whom we all derive our mitochondria; and that from this origin there were two main branches of the phylogenetic tree, one leading to six sub-Saharan mtDNA types and the other to everyone else.

Furthermore, if the descendants of that "mother" had bred with existing populations of archaic humans, ancient mtDNAs would have been incorporated into the gene pool. This has never been found in modern

BIFACE AXE

Acheulean tools (named for their first discovery at St Acheul in France) are characterized by having a long axis down which both sides have been worked to produce a crenulated cutting edge.

Taymyr ice sheet

Bering land bridge

NORTH AMERICA

Scandinavian ice sheet

Creswell Crags
Mladeč
Kostenki
EUROPE
Mezhirich
Cro Magnon
Bacho Kiro
Dar es-Soltane
Skhul
Qafzeh
Haua Fteah
Wadi Kubbaniya

Malaya Siya
Makarovo
Mal'ta
Kara-Bom
ASIA
Shuidonggou
Zhoukoudian
Zasaragi
Xiachuan
Nogawa
Sokchang-ni

AFRICA

Omo
Enkapune ya Muto
Katanda
Laetoli

Tabon
Niah Cave
Sunda land bridge
Bobangara
Wadjak
Sahul land bridge

AUSTRALIA

Apollo 11
Border Cave
Nelson Bay Cave
Klasies River Mouth

Devil's Lair
Lake Mungo
Kow Swamp
Bluff Rockshelter

modern human site
150,000–100,000 years ago
100,000–50,000 years ago
50,000–35,000 years ago
35,000–15,000 years ago

spread of modern humans
coastline, 18,000 years ago
ice cap, 18,000 years ago

vegetation zones, 18,000 years ago
semi-desert/desert
frozen steppe
savanna/grassland
temperate woodland/forest
tropical forest

PLEISTOCENE

samples. The implication of this is that when modern humans spread out of Africa they replaced, rather than interbred with, the established *Homo erectus* populations that had preceded them.

As might be expected, the "out of Africa" model faced considerable opposition and criticism of some of its techniques. The major alternative—and opposite—viewpoint is the theory of multiregional evolution, by which *Homo sapiens* is supposed to have emerged throughout the Old World (Africa, Europe, and Asia) through the gradual evolutionary change of the established archaic populations, without either significant migration or replacement of existing populations taking place.

Variations on the multiregional model are more extreme. One suggests a scenario of geographically isolated groups dating back a million years, implying a deep genetic division between them, which is supposed to account for the range of human diversity known as "races." Although human populations have been isolated at times, this has never been for long enough to create genetic barriers to mating. Despite persistent attempts to categorize humans by race, this classification has no biological value: the fact is that more genetic variation exists within a population than between populations.

A multiregional mode of evolution, with a common origin in *Homo erectus* ancestors that left Africa at least a million years ago, would display extensive mtDNA variation in modern populations. This has not been shown to be the case. By contrast, and although it remains controversial, the single-origin model has been confirmed again and again by newer studies corroborating the genetic data. The fossil evidence is also strong.

AN EARLY OCCUPATION

Meadowcroft is a rock-shelter located about 30mi(48km) south of Pittsburgh, Pennsylvania. Its deep deposits contain 11 major layers, and radiocarbon dates indicate that the site was occupied intermittently from at least 14,000 (and perhaps even 17,000) years ago until about 250 years ago, giving it the longest history of occupation in the New World.

Skulls from Omo (Ethiopia), Laetoli (Tanzania), Border Cave, and Klasies River Mouth (both South Africa), dating to between 150,000 and 100,000 years ago, are all those of recognizably modern humans (*H. sapiens*), albeit with some archaic traits. Similar remains have been discovered in caves at Qafzeh and Skhul, in Israel; these specimens have high, short braincases with fairly vertical foreheads, only slight brow ridges, well-defined chins, and a cranial capacity of about 95in^3 (1550cc)—all modern features. They have been dated to 100,000–90,000 years ago and provide the earliest known evidence for the presence of modern humans outside Africa.

The oldest known *Homo* fossils in Europe were found at Atapuerca in northern Spain and date from 800,000 years ago. These fossils show marked differences with their African contemporaries and were given the name *H. heidelbergensis* by paleoanthropologists (after a

THE FIRST AMERICANS

(Below) The question of when the first humans crossed the Bering landbridge and entered North America is hotly debated. The most widely accepted view is that a passage was found between the Cordilleran ice sheet in the east and the Laurentide in the west, about 13,000 years ago. However, some sites south of the ice, like Meadowcroft Rockshelter in Pennsylvania, have produced earlier settlement dates, and there are claims that some sites in Brazil and Chile are as much as 33,000 years old.

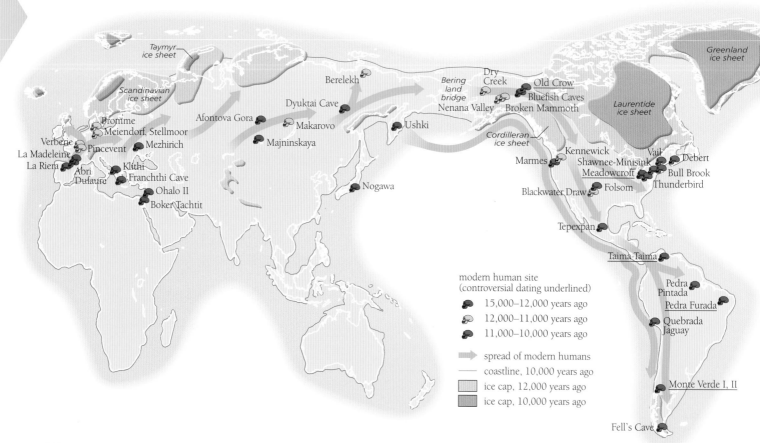

modern human site
(controversial dating underlined)

- 15,000–12,000 years ago
- 12,000–11,000 years ago
- 11,000–10,000 years ago

→ spread of modern humans
— coastline, 10,000 years ago
□ ice cap, 12,000 years ago
▨ ice cap, 10,000 years ago

specimen discovered at Mauer, Germany). Over thousands of years, archaic humans in Europe developed characteristic features, termed Neandertal (after the Neander Valley, Germany, where they were first identi-fied), many of which reflect adaptation to cold climates. Typically, they possess a powerful, stocky, short-limbed frame with a large projecting face and robust teeth. When *Homo sapiens* subsequently spread into Europe from the Afro–Asian region about 40,000 years ago, they displaced the local Neandertal populations, which had disappeared by 30,000 years ago. Modern humans reached southern Siberia by 40,000–35,000 years ago and Australia about 32,750 years ago. From Siberia they migrated across the Bering land-bridge into America.

DURING the period of the last major glacial, humans occupied much of the Earth. Unlike earlier forms of humankind they responded to the particular demands of their environment primarily through cultural and tech-nological adaptation. One of these behavioral adapta-tions was complex coopera-tive hunting of very big game using weapons. The expansion of human popu-lations and their effective-ness as hunters has been blamed, to a great extent, for the disappearance, between 12,000 and 10,000 years ago, of many species, particular-ly large herbivores, on many continents. This idea has become known as the "overkill hypothesis."

There is a coincidence between the spread of modern humans across the globe and the disappearance of large herbivores.

In North America 33 genera (about 73 percent) of large terrestrial mammals died out, including all the pro-boscideans (mammoths, elephants, and mastodonts), many horses, all camels, tapirs, and the huge, heavily-armored glyptodonts and the immense ground sloths. Many deer species became extinct, as did many of the predators that lived off this abundant source of meat. They included the North American lion (often referred to as the American giant jaguar), the saber-toothed cat *Smilodon,* and the scimitar-toothed cats such as *Homotherium.* In Australia five species of marsupials vanished, as did the giant monitor lizard; in South America 46 genera died out, including many of the forms characteristic of this continent such as litopterns, noto-ungulates, and edentates. Losses in Europe were less severe although woolly rhino, giant deer, and woolly mammoth became extinct; others—such as hippo and hyena—simply contracted their geographical ranges.

It is suggested that large animals are particularly at risk from intensive hunting because of their extremely long gestation period and slow growth, which means they attain sexual maturity only after many years. Losses would inevitably have greater effects than on small, fast-growing, highly fecund animals. Humans' excessive hunting of these animals eventually reached a point at which the population was not of viable breeding size, as removal of individuals exceeded the birth rate.

HALL OF THE BULLS

The modern humans that reached Europe are often referred to as Cro-Magnon people; they brought with them a char-acteristic array of advanced tools and covered the interiors of caves in France and northern Spain with equally characteristic paintings. The Hall of the Bulls in the cave of Lascaux, in south central France, is a significant source of archaeological, anthropological and semi-paleontological data. It provides a record of some of the animals which shared these environ-ments with ancient humans about 16—17,000 years ago and corrobo-rates fossil finding of aurochs (wild cattle), bison, horses, ibex and reindeer that come from this area and date from this time. A wounded bison suggests that these ancient humans hunted large-bodied animals as part of their food intake.

Supporters of the overkill hypothesis point out the good correlation between the spread of human popula-tions and the extinctions, and that the only prey to suffer were the large-bodied animals that were attractive food sources. That the big game of Africa were not affected to the extent of similar populations elsewhere is explained by the fact that humans had evolved in Africa alongside their prey species, who had time to adapt to the threat they presented. In those parts of the world where humans were new, animals didn't recognize them as predators and were easy targets. In addition, humans are "switching predators": that is, as soon as one source of food disappears, they will turn to another.

There is also circumstantial evidence that climate changes, at least in Europe, had little effect on many animals and that earlier glacial retreats did not necessari-ly cause extinctions. Unfortunately, a lack of archaeologi-cal evidence of kill sites, and the fact that humans entered Australia (and possibly also America) long before the bulk of the extinctions there took place, reduces the strength of the overkill hypothesis as the main cause of extinctions of large Pleistocene herbivores. Recent re-analysis of the dating and geographical spread of the extinctions in North America suggests that what happened was in reverse from the overkill hypothesis. Although the fact that non-prey species also died out is an argument for the climate-change model, it also supports a combined theory or "keystone species" hypothesis. According to this, the disappearance of large herbivores, as a result of hunting, had an impact on habitats that was detrimental to smaller animals. It was probably a combination of environmental factors and hunting that led to the demise of many of the larger animals.

PLEISTOCENE

ONE OF the richest sources of Pleistocene mammal remains is the Rancho La Brea tar pits, on the outskirts of modern-day Los Angeles, where, mistaking them for water holes, unwary animals became mired in the sticky deposits. The quantity and diversity of animals found there is astonishing; the remains of over 4000 individuals of 40 different mammal species and over 100 species of birds have been recovered. The ancient landscape, about 2 million years ago, would have been similar in many respects to open grassland habitats of today. Grasses and broadleaf trees flourished, along with shrublike plants such as the still-living Californian sage. Plants and animals showed a diverse warm-weather habitat. The largest animals there were the huge mammoths, some of which displayed immense upwardly-curving tusks. These were an indirect indicator of the high nutrient load of this ecosystem, since tusks of such size require considerable upkeep in the form of food intake. Mammoths are not, however, the most numerous large herbivores at the tar pits of La Brea; these were the high-withered bison. Bison fossils are found in such abundance that their dentition has a produced some startling information regarding the age structure of their community. Dental anatomy shows that individuals are always of a distinct age: 1 or 2 or 3 years old, but not 1.5 or 2.5 years old. This, and the patterns of wear on their teeth, show that the bison were coming to La Brea only at specific times of year, and were absent during other periods. In other words, the bison were migrating just as they did until recently, before they were almost wiped out in the last 150 years or so.

Animals struggling in the treacherous tar deposits at Rancho La Brea would have attracted the attention of predators and scavengers.

1 *Bison antiquus*
2 *Canis dirus* (dire wolf)
3 *Teratornis* (vulture)
4 *Smilodon* (saber-toothed cat)
5 Imperial mammoth
6 Coyote
7 Heron

PLEISTOCENE

One of the unusual things about the fossil assemblage at Rancho La Brea is the disproportionate number of carnivores, which make up nearly 90 percent of individuals. Most abundantly represented is the dire wolf, closely followed by the saber-toothed cat *Smilodon*, of which thousands of partial specimens and three fully preserved specimens are known. Preservation of animals at different growth stages demonstrate how the little canine "milk teeth" of the cubs were replaced by the immense 9in (23cm) curving "sabers" of the adults. Along with *Smilodon* was the giant short-faced running bear, with meat-shearing teeth and powerful jaw muscles; it stood 6ft (1.8m) tall at the shoulder and probably weighed close to a ton. The kills of these powerful predators would have been observed by the vulture *Teratornis* ("frightful bird"), which had a 13ft (4m) wingspan. Even larger vultures are known from elsewhere in California.

PLEISTOCENE

THE EVOLUTION OF HUMANS

Homo *sapiens*—anatomically modern humans— are defined by a thick cranial vault; a foramen magnum (the hole where the spinal column joins) situated directly beneath the skull; a reduced "snout"; a large brain; small cheek teeth, a well-developed chin; and no eyebrow ridges. *Homo sapiens* may claim a lineage of hominini ancestors, a small, specialized group within the primate subfamily Homininae, which separated around 5 million years ago from the gorillas and chimps that form the rest of the family Hominidae.

Cladistics show australopithecines to be the sister group to *Homo*. They had been assigned to one genus, *Australopithecus*, but in 1990 new finds showed that there are three genera: *Ardipithecus ramidus*, the oldest known hominid at 4.4 million years old; *Australopithecus;* and *Paranthropus*. *Ardipithecus*, from Ethiopia, has relatively large canine teeth and narrow molars with thin enamel, indicating a diet of leaves and fruit. These teeth are more hominine than any of the living apes. The foramen magnum is placed forward, showing that *Ardipithecus* must have walked on two feet.

Australopithecus anamensis was found by Maeve Leakey in sediments of 4.1–3.9 mya near Lake Turkana in Kenya. Anatomically closest to *Ardipithecus*, it is considered the most primitive australopithecine. *A. afarensis* remains have come from the Hadar region of Ethiopia, the most celebrated of which was the 40-percent-complete skeleton of a young female, nicknamed "Lucy," dating to about 3.18 million years ago. Later australopithecines include *A. africanus* and *P. robustus* from southern Africa; *P. boisei* and *P. aethiopicus* from eastern

HUMAN ANCESTORS

The human family tree appears rather sparse in that it depicts a taxonomically restricted view of a single group of fossils. Furthermore, humans are not very diverse and their fossils are extremely rare. Because the evidence is so meagre, any tree is, at best, a summary and many of the connections are open to debate. The main division is between the small-brained australopithecines and the large-brained, fully bipedal *Homo*.

1 Divergence from chimpanzees
2 Divergence and radiation of australopithecines
3 Appearance of robust australopithecines
4 Appearance of early *Homo* species
5 Acquisition of obligate bipedalism
6 Divergence of *Homo sapiens* subspecies
7 Coexistence of Neandertals and modern humans
8 Extinction or subsumation of Neandertals into the modern human gene pool

BEST FOOT FORWARD

Along with intelligence and manual dexterity, the most characteristic feature of humans is the bipedal form of locomotion, an important adaptation to a savannah environment (left). In order to pull the upper torso into an upright posture and position it directly over the legs, the buttock muscles must be large enough to lever the weight of the body into such a position. Accordingly, human buttocks are much bigger than those of the chimpanzee and gorilla. These muscular arrangements are reflected in the comparative anatomy of hominid pelvises; the human pelvis is more bowl-shaped than that of other hominids.

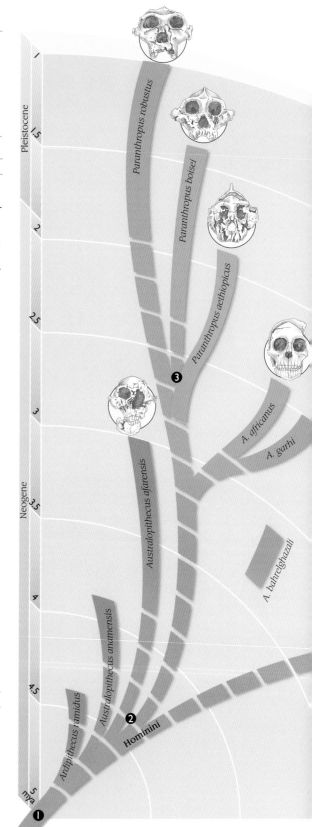

Africa. The *Paranthropus* species are, in many respects, the most interesting of all the australopithecines. These "robust" forms have broad shield-like faces, huge molar teeth, and a heavy ridge along the cranial midline that anchored powerful jaw muscles adapted to a tough diet.

There is great controversy regarding the recognition of several species of *Homo*. Modern anatomical and biomechanical studies show that the spectrum of variations among modern humans is greater than that of these fossil species. Many scientists regard the seven species of *Homo* as merely geographical variants of the normal anatomical variations that are usually encountered in any

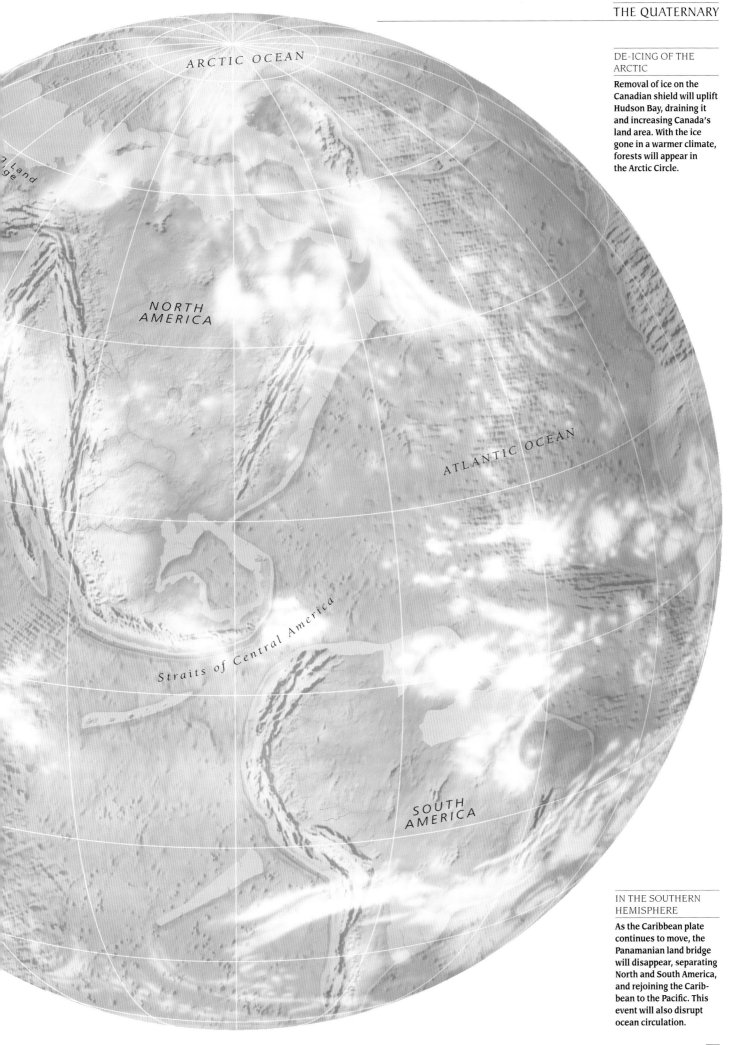

ARCTIC OCEAN

n Land
ge

NORTH
AMERICA

ATLANTIC OCEAN

Straits of Central America

SOUTH
AMERICA

DE-ICING OF THE ARCTIC

Removal of ice on the Canadian shield will uplift Hudson Bay, draining it and increasing Canada's land area. With the ice gone in a warmer climate, forests will appear in the Arctic Circle.

IN THE SOUTHERN HEMISPHERE

As the Caribbean plate continues to move, the Panamanian land bridge will disappear, separating North and South America, and rejoining the Caribbean to the Pacific. This event will also disrupt ocean circulation.

HOLOCENE

Western Europe will be pushed east as the Atlantic widens, infilling the North Sea with sediment from the delta of the Rhine River. The Mediterranean should disappear entirely as Africa pushes north, and be covered by a towering, intensely folded mountain range like the Himalayas. Similarly, eastern Africa splits off and drifts north, crashing into Iran and raising yet another tall mountain chain. The part of Africa between the rift valleys will become a partially submerged plateau, with isolated parts appearing as "islands" like the present-day Seychelles. The presently youthful Himalayas will still be tall, as India continues to be subducted beneath the edge of the Asian continent. Higher sea levels will flood the Ganges and Indus plains, making India appear more of a peninsula. Australia has moved so far north that it has collided with southeast Asia, raising steep mountains between them, and completing the great landmass of Africa-Eurasia, a new supercontinent. However, supercontinents do not remain so, and this one is already fragmenting into rift valleys along Lake Baikal and Lake Balkhas in central Asia.

Climates and biomes will have undergone many changes during this time. Antarctica, for instance, will have turned green as its ice melted, and forest will once again cover a great belt across northern Europe. Many species will have died, however, as continents moved through different zones, and in and out of ice ages.

TEMPERATURE increases associated with the current interglacial period have been supplemented by a rise that began in the early part of the nineteenth century as the Industrial Revolution gained momentum and atmospheric pollutants—chiefly in the form of burned coal products—began to fill the air with an extra load of carbon dioxide gas. In addition, there seems to be a gradual increase in global temperatures that began sometime in the mid-nineteenth century, although the upward trend was interrupted between 1940 and 1970 and shows as a plateau on a temperature graph. In the past 30 years, however, temperatures have risen sharply.

"Global warming" is a normal aspect of an interglacial episode, but it has clearly been accelerated by the burning of fossil fuels.

There is much concern about the "greenhouse effect," by which the increased quantity of carbon dioxide gas in the atmosphere acts as a blanket in the atmosphere, keeping heat close to Earth and raising the global temperature. Some thinkers argue that the greenhouse effect may be a natural phenomenon in the development of our planet and nothing to do with the appearance of industrialized societies, while others are adamant that the only possible explanation is human activity and the increased level of carbon dioxide gas in the atmosphere.

Higher temperatures and levels of atmospheric carbon dioxide would probably be very beneficial for plants, but not for humans. It is generally assumed that global warming would melt the polar ice caps, causing sea levels to rise and flood coastal regions around the world, with a catastrophic loss of living space and farmland. Ice is less dense than water, and so the melting of sea ice—which is already part of the sea's volume—would not raise sea levels; it would lower them very slightly. If continental ice at the poles melted, it would be a totally different matter. Continental ice is not a part of the sea volume. Melting of the Antarctic ice cap would add enormously to the sea volumes and sea levels might increase by approximately 200ft (60m). On the other hand, global warming might actually expand the Antarctic ice cap: with higher temperatures, evaporation of seawater increases, building up more water vapor in the atmosphere and increasing snowfall. It is difficult to predict which outcome is more likely.

If sea levels do rise with global warming, they will bring major regional climate changes. Areas with high rainfall, such as northwestern North America, will become wetter, again due to increased evaporation of warmer water. Tropical regions will experience more severe monsoons, because these storms are drawn inland by warm landmasses, which will have become warmer. Finally, higher temperatures will exacerbate the tendency to dryness in the interiors of large continents, compounding the expansion of desert as trees are cut down and erosion wears out the land.

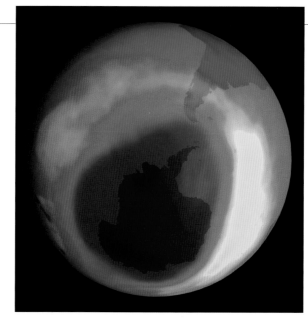

THE OZONE HOLE

A hole in the ozone layer, which blocks harmful ultraviolet radiation from the Sun, is visible over Antarctica each spring and seems to be growing; by 2000 it was the size of the United States.

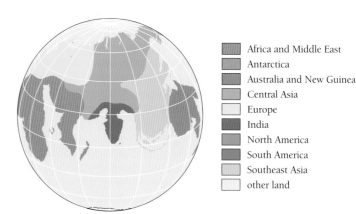

■	Africa and Middle East
■	Antarctica
■	Australia and New Guinea
■	Central Asia
■	Europe
■	India
■	North America
■	South America
■	Southeast Asia
■	other land

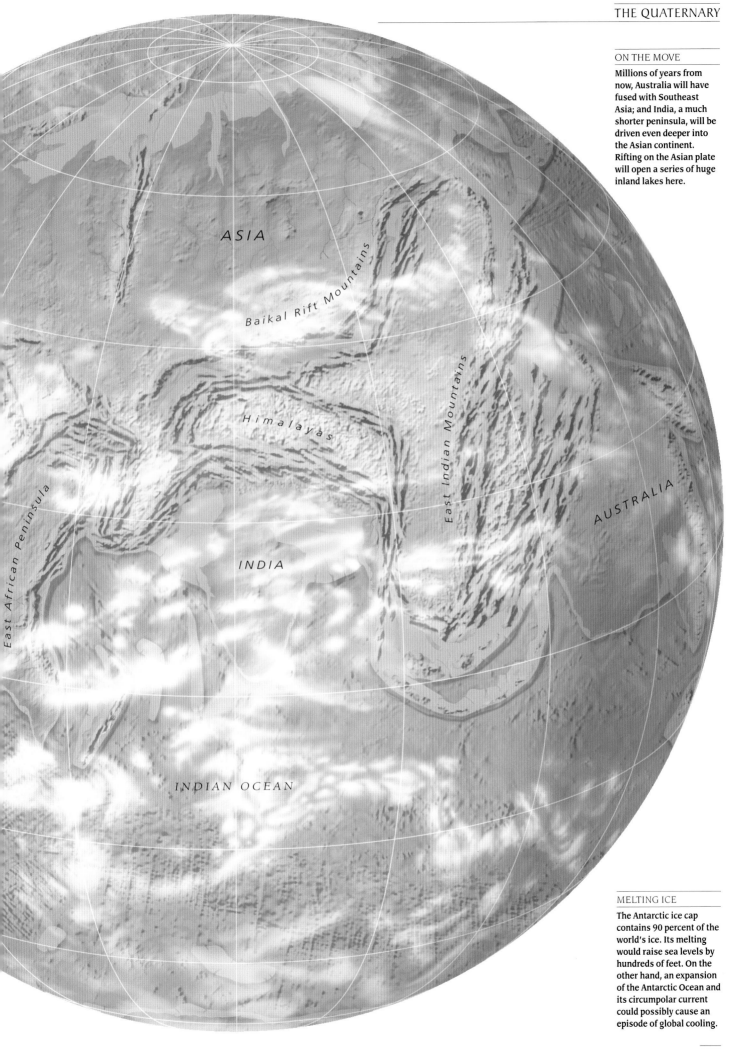

ASIA

Baikal Rift Mountains

Himalayas

East Indian Mountains

East African Peninsula

AUSTRALIA

INDIA

INDIAN OCEAN

ON THE MOVE
Millions of years from
now, Australia will have
fused with Southeast
Asia; and India, a much
shorter peninsula, will be
driven even deeper into
the Asian continent.
Rifting on the Asian plate
will open a series of huge
inland lakes here.

HOLOCENE

MELTING ICE
The Antarctic ice cap
contains 90 percent of the
world's ice. Its melting
would raise sea levels by
hundreds of feet. On the
other hand, an expansion
of the Antarctic Ocean and
its circumpolar current
could possibly cause an
episode of global cooling.

If the greenhouse effect does not interfere with the onset of the next glacial interval, will there be another ice age? If so, then it is the northern hemisphere that would be most affected as ice sheets again descend on the major industrialized and food-producing regions of some of the wealthiest countries on the planet. Technology may become sufficiently developed to shield these nations from climate conditions; climate-proof research stations on Antarctica show that this is possible. Nevertheless, agriculture and life in western Europe, Scandinavia, Russia, and North America would suffer.

MODERN Africa provides a dramatic picture of tectonically-driven change in progress. The eastern part first attracted the attention of geologists in the 1920s because of its unique geomorphology, tectonic significance, and associated volcanic activity. Inland southern Africa is a classic flat plateau of wide topography and very high elevation. A large part of southern Africa is more than 6500ft (2000m) above sea level—although the Kalahari basin is lower than sea level—and this huge high expanse is characterized by an even more celebrated suite of impressive geological features that have drawn the attentions of geologists from most countries: the African Rift Valley system.

Signs of a future split in Africa are already apparent in the great Rift Valley of the east, a product of faulting.

The term "rift valley" was introduced in 1921 by J.W. Gregory, who recognized the rift valley of East Africa as a product of faulting. He defined the term as a long strip of land that had been let down between normal faults, or a parallel series of step faults. Parallel fault zones are about 30–50mi (50–80km) apart, and the down-dropped central portion is displaced down by about 10,000ft (3000m).

The Great African Rift Valley extends 1850mi (3000km) south from the end of the Red Sea through Ethiopia, Kenya, and Tanzania in a long fracture in the lithosphere of the African continental plate. A western arc goes from the southern border of Sudan and is occupied by Lakes Albert, Edward, Kivu, Tanganyika, Rukwa, and Malawi, which have filled some of the deepest rifts. One of the most famous landmarks of the region is Mount Kilimanjaro, rising about 19,000ft (5900m) above sea level.

Modern rift valleys are closely associated with uplifted crust that may be major lithospheric domes or long undulating upwarps. The width of rift valley faults (30–50mi/50–80km) approximates the thickness of the rifted crust. Rifts are usually bounded by major normal faults that may occur in "steps" and that produce frequent shallow earthquakes. Hot springs, associated with the high heat flow produced by the molten material rising from the mantle, may also be present. As mantle material rises the lithosphere bulges. Crust at the surface stretches and cracks open, ❸ forming the rift valleys.

Rifting in East Africa is part of a regional process that dates back to the Tertiary period, when Arabia began to separate from Africa. The East African rift is part of a

MOUNT KILIMANJARO

The volcanic mountain of Kilimanjaro is Africa's highest peak, formed during the Pleistocene by rifting in the Great Rift Valley of eastern Africa. This makes it a relatively young mountain in geological terms—and, typical of areas where rifting is occurring, Kilimanjaro was active until very recently.

❷

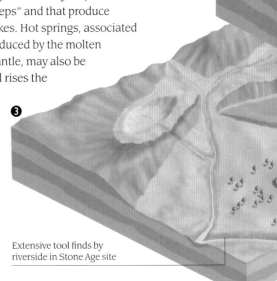

Extensive tool finds by riverside in Stone Age site

triple junction involving the Red Sea and the Gulf of Aden. It is very likely that the East African rift will become the zone at which African splits entirely, tens of millions of years hence. The Indian Ocean will flood in between the halves, forming a narrow extension of the ocean, as the Red Sea has flooded between Arabia and Africa.

VOLCANIC rocks in rift valley systems are usually extruded as outwardly-directed plateau flows. An example is the Trap Basalts that form much of the Ethiopian highlands. The total volume of rocks that are considered to be directly associated with the Eastern Rift in Kenya has been estimated to lie in the region of approximately 143,000 mi^3 (600,000 km^3), and there is an even greater volume in the Ethiopian Plateau. The oldest volcanic rocks in the Great Rift Valley system are from Ethiopia and have been dated to about 30 million years of age. These volcanic rocks have proved to be of immense impor- tance, since they can be

Volcanic rocks, which are abundant in the Great Rift Valley, have been important in the search for human origins in the region.

EARLY AND RECENT

(Below) An intriguing assortment of fossils comes from the Chemoigut Formation in Kenya, almost equally spaced between Olduvai Gorge and Lake Turkana. (1) Rocks of the formation were deposited 1.5 million years ago. Fossils of crocodiles and antelope are common in lakeshore sediments, and early Oldowan stone tools are present. (2) Strata of volcanic ash 250,000 years old covered the area in deep deposits, cut into by channels of rainwater. Hand axes found along these channels are of the advanced Acheulean type of the Late Paleolithic ("Stone Age"). (3) About 5000 years ago, the modern landscape of the rift valleys began to take shape. Bones and arti- facts suggest that a "factory" industry producing stone tools was operating by then.

assigned highly accurate dates by isotope analysis and radiometric geophysical techniques. Their dating has provided extremely well-defined stratigraphical chronologies for the fossils that have been found in asso- ciated sedimentary rocks of the region. The geological conditions in the rift valley—eruption, uplift, and erosion—have alternately preserved and exposed layers of life dating back several million years. In parts of the rift system in southern Ethiopia, northern Kenya, and northern Tanzania, hominid and other faunal remains have been discovered; and examples of the former have shed new light and provided fundamental data on the early origins of anatomically modern humans.

From Plio–Pleistocene times onward, the lakes that formed in the downfaulted regions of the rift valleys have attracted human habitation. Then as now, the water supply would have supported vegetation and drawn all wildlife in the vicinity to drink at the plentiful watering- holes. Such regions are always a gathering point for large animals. In addition, these lakes and smaller bodies of water were essentially geographical regions of sediment deposition. Sediments at the bottom of the lakes, left undisturbed by humans throughout thousands of years, preserved the remains of animals very well. The more quickly the bodies were buried, the better the preserva- tion. Layers of ash from erupting volcanoes in the valley added their protection. Modern Lake Turkana, which is more than 185mi (300km) long, and comparable in width to the English Channel, occupies an even larger basin in the generally arid eastern part of the Great Rift Valley. It formed in the vicinity of an ancient lake where the four-million-year old *Australop- ithecus anamensis,* the oldest australopithecine hominid, might have gone for water.

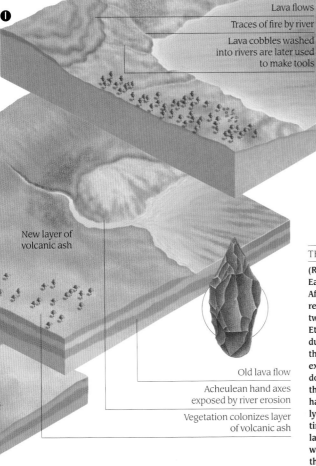

Lava flows
Traces of fire by river
Lava cobbles washed into rivers are later used to make tools

New layer of volcanic ash

Old lava flow
Acheulean hand axes exposed by river erosion
Vegetation colonizes layer of volcanic ash

THE RIFT VALLEY

(Right) Rifting of the Earth's crust in eastern Africa was the direct result of the formation of two huge domes, the Ethiopian and Kenyan, during the Neogene. As the overlying crust expanded above the domes, tension caused the rifts to form. They have become progressive- ly deeper since Miocene times, and have accumu- lated sediments and water. It was beside these lakes that the first hominids lived, and their fossils were preserved in the sediments.

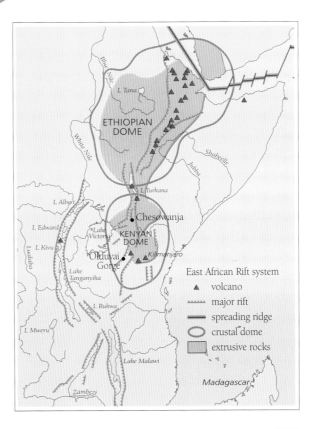

Blue Nile
L Tana
ETHIOPIAN DOME
White Nile
Shabeelle
Jubba
L Turkana
L Albert
Chesowanja
L Edward
Lake Victoria
KENYAN DOME
L Kivu
Kilimanjaro
Olduvai Gorge
Lualaba
Lake Tanganyika
L Rukwa
L Mweru
Lake Malawi
Madagascar
Zambezi

East African Rift system
▲ volcano
major rift
spreading ridge
○ crustal dome
extrusive rocks

HOLOCENE

CONTINUING tectonic activity in South America has produced a very different landscape. The Andes range stretches along the entire western margin of South America, from the Caribbean sea in the north to the Scotia sea in the far south—a distance of some 6200mi (10,000km), making the Andes the longest continuous mountain chain in the world. This great range is about 250mi (400km) wide and has a maximum elevation of about 23,000ft (7000m) above sea level. Cerro Aconcagua on the border between Chile and Argentina, at 22,834ft (6960m), is the highest mountain on the continent and in the entire western hemisphere.

Mountain-building is still in progress along the entire length of the west coast of South America. The Andes are one of the youngest ranges in the world.

Above all other features, the Andes are characterized by a huge number of very high and frequently very active volcanoes. Due to the bulge of the earth at the equator, the top of the immense volcano Cotopaxi in Ecuador, at a modest elevation of 19,347ft (5897m), is farther from the center of the earth than the top of Mount Everest, whose peak is 29,028ft (8848m) above sea level.

Most mountains are the product of collision between continents, but the Andes arose—and are still rising—as a result of subduction between the ocean plate of the southern Pacific and the continental plate of South America. A major tectonic system extends in a north-south trend along the western side of the whole of North, Central, and South America, beginning with the Juan de Fuca plate in the north, which is linked to mountain-building and earthquakes along the west coast of North America. This long coastal span is the demarcation zone of collision, where the ocean plates are being pushed beneath the edge of the continents by the spreading of the Pacific ocean ridges. The subducting Nazca ocean plate beneath South America is mainly responsible for the formation of the Andes Mountains; the Cocos and Antarctic plates to the north and south are also involved.

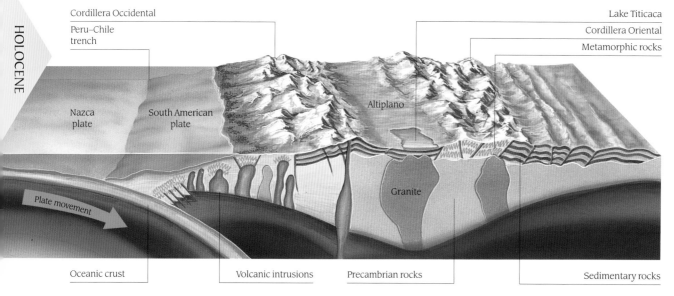

Cordillera Occidental

Peru–Chile trench

Lake Titicaca

Cordillera Oriental

Metamorphic rocks

Nazca plate

South American plate

Altiplano

Granite

Plate movement

Oceanic crust

Volcanic intrusions

Precambrian rocks

Sedimentary rocks

HOLOCENE

ANDEAN CORDILLERA

Subduction of the Nazca plate under the western edge of the South American plate is forming the Andes (left). Ocean crust and seawater are pushed to great depths, where they melt into a lighter magma than the surrounding material. The subducted material rises, pushing through the edge of the continental crust, and erupting as volcanoes. The range is unusually wide, because the plate subducts at a very low angle, causing the center of igneous activity to move inland by 125mi (200km).

As the southern plates curve downwards into the upper reaches of the asthenosphere beneath the western edge of the South American plate, the constituent minerals of these oceanic crustal portions are melted progressively at greater depth. These minerals are often less dense—and therefore lighter—than the minerals that make up the main portion of the mantle. As they melt, they rise up from deep in the mantle, collecting and pooling under the edge of the South American plate. This process extends far inland into the continent: about 500 miles (800km), which accounts for the unusual width of the Andes chain. Eventually these hot, liquid mantle materials, derived from the long-subducted Pacific oceanic crust, burst through the upper continental crust of the South American plate as volcanoes. These are still active throughout the entire cordillera system, from Mount St. Helens in the northwest to the Caribbean island of Montserrat, which spewed ash for a whole year in the late 1990s, through the length of the Andes to the volcanoes at the tip of Chile.

The Andes gave their name to a type of lava, andesite, which is the most commonly erupted type of lava in the region, where it is a major constituent of the Earth's upper crust. Andesite is a fine-grained igneous rock formed under certain conditions of magma crystallizing in the upper crust, and its presence is associated with highly explosive volcanic activity. Less viscous, slower-moving basalt lavas of the Hawaiian volcanoes tend to ooze rather than run.

Another geological peculiarity of the Andes is the presence of gigantic granite batholiths ("deep rocks")—immense globules of ancient magma that are located deep in the uppermost layer of the Earth's crust. This is the location of the fractional melting that produces andesite lavas. The Patagonian Batholith at the southern tip of South America is about 620 by 620mi (1000 by 1000km).On the other hand, the ancient seafloor fragments known as ophiolites, which are typical of mountains produced by continental collisions, are rare in the Andes, because most of the seafloor has been pushed under the edge of the continent rather than raised up. Dating of the Andean batholiths show that the events of subduction in this area began about 130 million years ago, sometime in the Early Cretaceous period. This was when the South Atlantic ocean ridge began to form, and the Pacific correspondingly began to shrink in size and the Pacific ocean floor started to converge with South America.

YOUNG MOUNTAINS

The Andes (above) are still forming, and they have had little time to erode. Both factors account for the exceptionally jagged shape of their numerous peaks.

MINERAL RICHES

The Andes are rich in minerals (right), both metallic and non-metallic, derived from geochemical changes that take place deep in the heart of the mountain belt. Uplift exposed the minerals to native Andeans, whose mined wealth was the target of European greed in the sixteenth century.

THE CARIBBEAN region formed as a segment of Pacific oceanic crust overrode a mid-portion of the Atlantic crust along a zone that now extends northward from South America east of the Lesser Antilles (from the Virgin Islands down to the coast of Venezeula). The small plate so formed is called the Caribbean plate. The Greater Antilles—Cuba, Jamaica, Puerto Rico, and Hispaniola—are a northeastern extension of the Andean mountain chain, which extends out the north coast of the continent and into the Caribbean sea; the Lesser Antilles are a volcanic chain that formed later, but both are intricately linked to the Andean system.

One of the last features of the modern globe to take shape, the Caribbean region is likely to be a short-lived phenomenon.

During glacial and interglacial episodes throughout the Pleistocene epoch, sea levels fell by as much as 420ft (130m) as the ice sheets removed water from the oceans. But sea level increased as the melting of the polar ice caps inundated much of this high terrain, leaving the islands, and parts of the old Cordilleran mountains, exposed. The Greater Antilles display distinctive karst landscapes formed by the erosion of huge limestone beds typical of the region.

Weathering of rock occurs particularly rapidly in the tropics, where the high humidity and rainfall help any overlying vegetation to decay faster. Sinkholes are another feature of karst landscapes, often measuring 60 to 180ft (100 to 300m) in diameter. Even larger depressions are known as cockpits, and these may be up to 125ft (200m) deep, as in the Cockpit Country of Jamaica. The karst landscapes there and in Cuba have been flooded by rising sea levels to produce a maze of underground caverns connected to the sea, decorated by

PANAMA
VENEZUELA
Nevado del Ruiz
Nevado de Huila
COLOMBIA
Nevado de Cumbal
Cotopaxi
ECUADOR
Sangay
BRAZIL
PERU
L. Titicaca
BOLIVIA
El Misti
Guallatiri
PARAGUAY
Antofall'a
Copiapó
CHILE
ARGENTINA
Aconcagua
Tupungatu
Tinguiririca
Azul
Villarrica
Osorno
Minchinmavida
Cerro Hudson
Cerro Lautaro
Monte Burney

▲ major volcanoes
▲▲ subduction zone
⬜ Andean cordillera

metallic ores
ⓒ copper
ⓖ gold
ⓘ iron
ⓢ silver
ⓣ tin
ⓞ other

◆ nonmetallic minerals

HOLOCENE

THE CARIBBEAN

(Below) The Caribbean formed over 125 million years at a subduction zone along the Pacific coast between North and South America. (1) About 100 mya, the subduction zone along the Farallon plate changed direction, and an island arc, the proto-Antilles (including Cuba, Puerto Rico, and Hispaniola), shifted east, beginning to consume the Atlantic plate. The northern arc collided with the Yucatán to form Nicaragua; the rest of Central America was a volcanic arc formed when the Antilles struck Cuba, creating a new subduction zone in the Farallon plate and isolating a piece that became the Caribbean plate. The Greater Antilles gradually separated. (2) By 10 mya, the Farallon plate had left two remnants, the Juan de Fuca and Cocos plates. About 3.5 mya, the land-bridge separated the Caribbean and Cocos plates. The Lesser Antilles arose where the Atlantic plate is being subducted under the Caribbean.

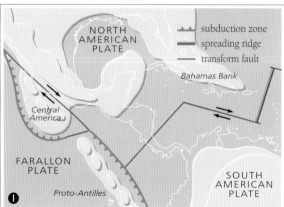

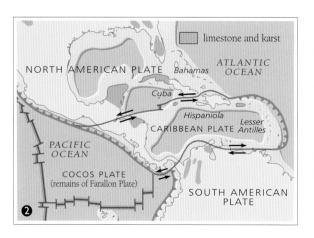

stalactites and stalagmites. A highly specialized light-hating biota has developed in these sea caves.

The development of slow-growing reefs in the Caribbean region shows how sea levels have changed in that area since the Pleistocene ice ages. Coral stops growing when exposed to air, and resumes when covered again with water, with younger segments growing on top of the earlier. As a result, a series of coral units are super-imposed upon each other rather like sedimentary strata, following the pattern of rising and falling sea levels, and can provide clues to the past offshore conditions of the Caribbean islands. These "stranded" reefs are especially good on the island of Barbados.

During glacial episodes of the Pleistocene epoch, tem-peratures in the Caribbean region were near the lower limit for the growth of coral reefs, inhibiting their spread. The effect of this temperature drop can be measured by comparing the numbers of coral species in the modern Caribbean with those of the western Pacific and the Indian Ocean: whereas there are about 60 surviving Caribbean species, there are nearly 600 in the Pacific and Indian oceans of today. Such changes reflect the increased melting of ice sheets about 18,000 to 12,000 years ago. This melting decelerated from about 12,000 to 10,000 years ago, and speeded up again afterwards. Rates of sea level rise were as high as 8.2ft (2.5m) per hundred years about 9500 years ago.

KARST ISLAND

(Above) Cuba is an island jammed onto a large limestone platform, the Bahamas Bank, about 55 million years ago. Its karst landscape resulted from the passage of rainwater through cracks in the limestone, where either the rock itself or the rocks that fill the fissures are solvent in rainwater. Intricately angled fissures, long-term erosion, and solution have an effect not seen in other geological forma-tions, producing relatively small but very sharp peaks. Karst derives its name from an outcrop in the former Yugoslavia.

SOUTH of the Caribbean is the huge delta of the Amazon River, which occupies the Atlantic side of the Andes chain. Here salt waters from the marine realm mix with fresh waters running off the higher regions inland to form enormous brackish-water environments. The rise and fall of sea levels during the past two million years has had profound effects upon life and geography in this region. Andean tectonic margins have also had a considerable effect on this region and its natural history. Active margins are the site of mountain building, but passive margins are low, hence the land slopes toward passive margins. In this case, the north part of South America is sloping down from the Andes mountain range to the Atlantic ocean on the eastern side of the continent. Much of the drainage in the Amazon basin is due simply to gravity, which has acted in concert with tectonic events in South America to produce the great expanse of Amazonian tropical rainforest that has been a feature throughout most of the Holocene epoch.

South America's Amazon Delta and its great rainforest are a side effect of regional tectonics and the fluctuating sea levels of the Pleistocene.

At first glance, a continental drainage map of the last 10,000 years might look somewhat haphazard, but in fact there are some consistent features: the largest rivers (those with the highest drainage) are always situated on sloping terrain where there is high rainfall. The Amazon discharges 3,675,691cu ft (104,083cu m) per second, approximately three times more than Africa's Congo River. The Amazon is a good model for an idealized continental river system, with a large collecting area in a nearby mountain belt, a trunk stream (river) flowing across a stable land platform or shield, and a passive margin to collect its output. The preglacial drainage of North America was also similar to this system, and what we see today is a feature that has only been fully developed in the Holocene: the river's main trunk arises in the Andes and flows across the downward-sloping passive margin of northern South America toward the Atlantic Ocean. More than a thousand smaller tributary rivers flow into the main trunk, many of which are major rivers in their own right. During the rainy season, water from high up in the Andes causes the main trunk to flood, covering up to 12mi (20km) of forest on each of its sides to a depth of up to 32ft (10m).

The glacial Amazon Delta was previously very much bigger because of the lowered sea levels that had caused an increase in the areal extent of the continent of South America. With most recent melting of the Arctic ice sheets about 10,000 years ago, this lower-lying area was submerged to form the current topography. Future rifting of the South American shield will cause further changes. The trunk of the main drainage system will be rerouted, draining new areas, and the new ecosystem will be most favorable to trees with water-resistant adaptations, such as mangroves.

Ecological changes have stimulated a number of innovations in the animals that inhabit the Amazon region. Perhaps the most unusual of these is found in the blind Amazonian river dolphin. Unlike marine dolphins, which have smoothly sloping foreheads, the forehead of the Amazonian dolphin slopes almost vertically down to its beak, and its beak is about twice as long. Its eyes are so reduced as to be almost non-existent: vision is useless in the muddy water of Amazonian estuaries, cluttered by millions of close-packed underwater tree roots. Echo-location is especially sharp in this curious dolphin, which can detect an object the size of a matchstick head in the huge volume of an Olympic swimming pool.

THE AMAZON BASIN

(Right) Amazon topography has always been connected with rising and falling sea levels. At times of glacial maxima, low sea levels exposed enormous areas of lowland tropical rainforest around the Amazon basin, whereas when the ice retreated, the high sea stands flooded these areas and diminished the areal extent of tropical lowland forests greatly, reducing them to isolated "islands." One product of submerging has been the the estuarine Amazonian rainforest, in which most of the plants are adapted to living partially submerged in brackish water.

lowland forest in the Amazon Basin

- at glacial minima
- at glacial maxima
- modern coastline and drainage
- land at glacial minima
- land revealed at glacial maxima
- Andean cordillera

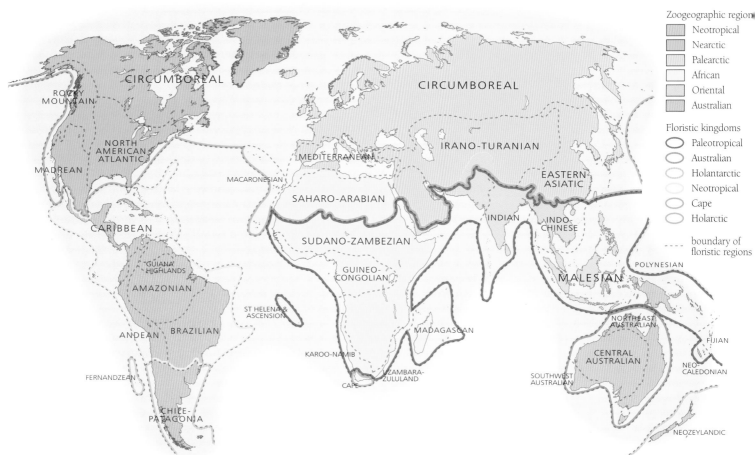

Zoogeographic region
▢ Neotropical
▢ Nearctic
▢ Palearctic
▢ African
▢ Oriental
▢ Australian

Floristic kingdoms
⬭ Paleotropical
⬭ Australian
⬭ Holantarctic
⬭ Neotropical
⬭ Cape
⬭ Holarctic
- - - boundary of floristic regions

Gʟᴀᴄɪᴀʟ cycles were responsible for the distribution and population densities of plants and animals as they stood at the beginning of the Holocene. Biogeographic realms are the divisions of these biotas, shaped by the locations of the continents—mostly stable throughout the Cenozoic—and the recent climates on each continent, which have changed dramatically as ice advanced and retreated around the world. Biogeographic realms are based on the most prevalent groups of plants and animals in each. Among fauna, these tend to be the larger vertebrate animals such as birds and mammals, while in the floral realm they are usually angiosperms (flowering plants).

Climate and the physical barriers of mountains and oceans have shaped the presence of plants and animals in the world.

There are six recognized modern biogeographic realms, based on the division into Old World and New World refined by the nineteenth-century naturalist Alfred Russel Wallace, which in turn corresponds to the separation of the continents in the Cenozoic. South America and Africa were isolated earlier than the other continents and have very distinctive mammalian faunas; that of India and southeast Asia, while similar to Africa's, has diverged enough to merit its own realm, the Oriental; and the Australian, the most recent, is unique in being dominated by marsupial mammals. The Nearctic and Palearctic realms (North America and Eurasia), where the Pleistocene ice sheets were most extensive, both lost many species and have remained impoverished.

A species may be found in only one place (endemic), in all places (pandemic), or in a variety of places (cosmopolitan). Rodents are pandemic; they are present in all biogeographic realms. Marsupials are endemic, found in the Neotropics and Australia. Perissodactyls and artiodactyls (hoofed mammals), elephants, carnivores, and lagomorphs (rabbits) are cosmopolitan, living everywhere except Australia; as are primates, which originally occurred in all but the Nearctic and Australian realms.

Animals and plants that survive in icy conditions tend to be drought-resistant, because frozen water is not available. Such plants grow too slowly and sparsely to support large herbivores. Animals in this ecosystem must

BIOGEOGRAPHY

(Above) **The biogeographic realms of the Earth are characterized by distinctive plants and animals. Temperature is the most important factor controlling what lives in each realm: more species, with greater diversity, live in tropical climates than in cold ones. Barriers imposed by land (such as mountains) and water also shape biogeography. Both climate and barriers may change over time.**

The European starling (left) is remarkably adaptive to new environments, which has not been good news for the native birds of North America, which it displaced. The starling was introduced into New York state in 1905. In just 50 years (below), it could be found in almost every part of North America, ousting the American bluebird as it spread. Huge flocks of these noisy birds are now a familiar sight from northern Canada down to the Gulf of Mexico and across to the Pacific coast.

AUSTRALIAN REALM

(Below) The Australian biogeographic realm is particularly well defined. Its fauna is dominated by large lizards and marsupial mammals. Placental mammals have, for the most part, been introduced by Europeans in the last 200 years. The Australian flora is also distinctive for its high proportion of sclerophyllous plants, which have thick waxy cuticles to resist prolonged drought.

be able to extract water efficiently from what little plant food there is. Many of them hibernate, minimizing their energy requirements when the least food is available. In temperate grasslands, in contrast, abundant food supports large animals with high energy requirements. Both circumboreal and grassland ecosystems are present in the Nearctic and Palearctic realms, which are subdivided into biomes. Tropical biomes have the greatest number and diversity of species: there are 137 families of flowering plants in South America, compared with 94 in North America. One characteristic of the tropics is the stability of temperature and rainfall, which do not vary much throughout the year. Tropical species are the first to disappear when conditions change.

THE CHANGES wrought by plate movements and glacial cycles are now being compounded by human interference, as we (often unwittingly) distribute plants and animals into entirely different regions. Australia's isolation 50 million years ago contributed to the evolution of its unique marsupial mammals such as the kangaroo. Placental mammals have only been introduced by humans in relatively recent times.

Biogeographic realms of the future are being shaped today by human activity, which is having a profound effect on biomes.

Because of its long-standing isolation there are still very few big mammalian carnivores in Australia. In the past, considerable numbers of large reptilian predators in the dry climate tended to exclude sizeable mammalian predators. This was simply due to low primary productivity, which restricted the numbers of large herbivores and carnivores (which feed on herbivores). The introduction of domestic dogs about eight thousand years ago displaced two mammalian carnivores, the Tasmanian devil and the Tasmanian wolf (thylacine), from the mainland to the Tasmanian island, and contributed to the extinction of the Tasmanian wolf.

Introduced species may overrun an ecosystem as well as displacing its native inhabitants. The colonization of Australia and New Zealand by Europeans provides more examples. Australia had no hoofed grazing animals such as sheep and cattle, because its native grasses were more fragile than those in the North American and European biomes. The import of these animals quickly wrecked Australia's grasslands. Replanting the region with tougher imported grasses did no good, because they

HOLOCENE

were poorly adapted to the hot, dry climate with its summer bush fires. Rabbits, another introduced species, experienced a population boom and depleted the ground cover still further—to the extent that a viral disease, myxomatosis, was deliberately introduced in the 1950s to reduce their numbers. Yet more destruction of the original Holocene ecosystem was caused by the introduction of decorative plants such as the prickly pear cactus, which invaded vast stretches of open plains and water holes in the tropical north, taking over 65 million acres (25 million hectares) before an effective biological control for it was found.

MODERN biogeographic realms are complicated by relict populations: lingering groups of animals or plants that were once widespread in an area but whose numbers have now greatly diminished. In Europe, these relicts may be directly attributed to the effects of the Pleistocene glaciations. Many species that were widely distributed in the past now exist in only a few "islands" of favorable climate or physical conditions. They are not necessarily species with long evolutionary histories, because some climatic changes took place relatively recently. Plants and animals that lived on or near to ice sheets would have had to be adapted to the freezing conditions, which extended almost as far south as the

Mountaintops and other isolated habitats, such as caves, are ideal refuges for species that once had a much wider distribution.

Mediterranean. Since the beginning of the Holocene interglacial, cold-adapted animals have been restricted to the very coldest areas of Europe. These are invariably at higher altitudes among mountain ranges, and consequently the same animals that were once found close to the subtropical Mediterranean have moved north, up into Scandinavia, Iceland, and Scotland. Certain species have become extinct in northern areas and are represented in European biotas as glacial relict populations of extremely restricted distributions. A good example is the tiny primitive insect known as the springtail. *Tetracanthella arctica* is a dark blue springtail, about 0.05 inches (1.5 mm) long, that lives in the surface layers of the soil and in clumps of moss and lichens, where it feeds on dead plant tissues and fungi. It is particularly common in coastal Greenland, all of Spitsbergen and Iceland, and a few areas of northern Canada. But outside these truly arctic regions it is known to occur in only two regions: the Pyrenees Mountains between Spain and France, and the Tatra Mountains on the borders of Poland and Czechoslovakia, with very isolated finds in the Carpathian Mountains to the east in Romania. In these mountain ranges, *Tetracanthella arctica* is found

STRAWBERRY TREE

(Left) The strawberry tree shows the swollen red fruit from which it derives its name. This tree has a broad distribution across southern Europe and a single relict population in Ireland. It was probably the result of distribution during the last Ice Age of the Pleistocene.

- ○ Norwegian mugwort
- springtail
- strawberry tree
- magnolia
- gorilla
- ◯ dung beetle
- ◯ tulip tree

GORILLAS

Gorillas (left), strictly speaking, are not relict populations but disjuncts: descended from a common ancestor, they split into highland and lowland groups on either side of the range of the ancestor, which then became extinct. Today the highlanders and lowlanders are kept separate by the geographical barrier of the Zaire (Congo) River.

RELICT SPECIES

(Above) Norwegian mugwort, magnolias, and dung beetles as well as springtails, tulip, and strawberry trees all appear in more than one population, but in far-flung locations. All of these are old species that originated when the continents were still close enough together to permit distribution across land areas that have now drifted far apart.

HOLOCENE

SPRINGTAILS

SPRINGTAILS

(Right) Springtails are primitive arthropods with a long evolutionary history. Modern populations are mostly found in Greenland, Iceland, and Spitsbergen, with two relict populations in continental Europe.

arctica has not been discovered at high altitudes in the Alps, much closer to home, where suitable conditions prevail. It may be that it once occupied a glacial relict position there but has died out in very recent times.

A plant example of a glacial relict is the Norwegian mugwort, *Artemisia norvegica*, a small Alpine plant that is now restricted to two areas of Scotland and very small regions of Norway and the Ural Mountains in Russia. During the last glacial period and immediately afterwards, the Norwegian mugwort was very widespread, but it was heavily curtailed as the post-glacial European forests expanded. It is almost certain that there are several hundred glacial relicts of this sort from both the plant world and animal world existing in Eurasia, including many forms that—in contrast to the springtails—could have traveled more easily.

One example of organisms that fit this criterion is the mountain or varying hare *Lepus timidus*. The varying hare derives its name from its appearance, which varies with the seasons: its fur is white in winter but tinged with blue for the rest of the year. It has a circumboreal distribution (Scandinavia, northern Japan, Siberia, and Northern Canada) and is closely related to the common Brown hare. The varying hare is found as relict populations in Ireland, the Southern Pennines in England—climates that are not particularly cold—and as a small population in the Alps. A factor in its distribution is that it is apparently poor at competing directly with the brown hare but is much better adapted to the cold. Yet another curious relict is the dung beetle *Aphodius holderi*. This big beetle is found in the high Tibetan plateau and as recent fossils dating from the middle of the last glaciation from a gravel pit in southern England, not far from the river Thames.

in arctic and subarctic conditions at altitudes of around 6600ft (2000m). It is clear that it would have been impossible for this tiny insect to colonize the Pyrenees and Tatras from its main habitation so much further to the north, from which it is separated by open water and large expanses of warm low-lying land—conditions that would be fatal to this animal. Springtails are quickly killed by low humidity as well as higher ambient temperatures; and it is extremely unlikely that springtails were transported from such inhospitable regions to such high isolated areas by humans or their domesticated livestock.

The most likely explanation for the presence of springtails in the Pyrenees and the Tatras is that they represent remnants of a much wider distribution in Europe during the Ice Ages. This animal expanded as Pleistocene ice sheets covered the continent, and survived only in a few pockets of suitable environmental conditions after the ice retreated north. It is surprising that little *Tetracanthella*

TULIP TREES

(Right) There are only two species of tulip tree known today, found in two widely separated locations: eastern North America and southeast Asia. This scattering suggests that they once had global distribution and the modern populations are relicts.

HOLOCENE

SUCCESS in evolutionary terms is defined differently for different species: mammals that survive for a million years may be said to be successful, whereas the average for marine species is 10 million years. In general, animals with highly specialized adaptations tend to go extinct most easily, whereas those that occupy a more general niche are often the ones that survive mass extinctions. Rodents, for instance, can feed on absolutely anything and live anywhere, as can canids, which are the most general of carnivores. Saber teeth, however, while unsurpassed for preying on large, slow, thick-skinned herbivores, are evolutionarily fragile: when large herbivores die out due to climatic and floral changes, saber-toothed carnivores cannot adapt quickly to different prey, and so they are also lost. Nevertheless, the repeated recurrence of saber teeth among Cenozoic carnivores shows that it is an excellent design for this specialized lifestyle.

Whether or not a species is successful in the long term depends to a large extent on environmental factors, including the presence of other animals.

Other species are important to the cycle of population growth of a given animal (or plant). There is a close relationship between predator and prey in any population. Rodent populations grew explosively from the Oligocene onward as global cooling caused savannahs and prairies to spread. The rodents flourished in these habitats but were of no use to big cats and other large carnivores, which could not get at them in their burrows. Into this niche came a new line of small, long-bodied, highly specialized carnivores. Nearly all of them belonged to the mustelid family of true mammalian carnivores. They included weasels, stoats, otters, badgers, skunks, and wolverines, but it was only the stoats and weasels that developed the underground burrowing mode of hunting. They have been highly successful for millions of years.

One mustelid, the black-footed ferret of North America, was once abundant due to the huge populations of the fat burrowing rodents known as prairie dogs which were present on the North American plains of the last 200 years. Throughout the eighteenth and nineteenth centuries, prairie dogs were hunted until their population crashed, and the black-footed ferrets promptly went into a sharp decline. Demand for their fur reduced their numbers further until they were thought to be totally extinct in the wild. Then, in 1986, ten individuals were found in Wyoming, and captive breeding was begun.

Another mustelid, the stone marten of Europe, has survived overhunting and colonized urban areas, where it is known to eat rubber parts from car engines.

BLACKFOOT FERRET
(Left) This mustelid's dependence on prairie dog prey nearly caused its extinction when prairie dog populations on the North American plains were nearly wiped out.

CAPE VULTURES
(Below) Vultures feed on dead carcasses but need large carnivores to open them. Vulture populations in South Africa nearly crashed when farmers all but eliminated hyenas, which crack bones into splinters small enough for vultures to use as food for their chicks. The chicks lacked calcium and consequently failed to develop.

EVOLUTIONARY fragility seems to be as much a function of behavioral adaptation as of anatomical suitability, but this is very difficult to infer from fossils. The stone marten and the black-footed ferret, though successful, illustrate vulnerability at the hands of humans. Another familiar example of this is the cheetah, *Acinonyx jubata.* Its oldest fossils are about 3.5–3.0 million years old and come from southeastern Africa. The modern species is restricted to Africa, though fossils are found in Asia and the Middle East as well. Living cheetahs feed exclusively on Thomson's gazelle, which has profound implications for their evolutionary prospects.

> *In the well-known case of the cheetah, being the fastest predator of all time is no guarantee of being able to eat and survive.*

In Europe the cheetah was present until as recently as 500,000 years ago. Fossils show it to have been much larger than the modern African version—an adaptation to the colder climate—and possibly even faster, if it was not slowed down by its greater weight. It is curious that the skeleton of the living Himalayan snow leopard, with its small head and slim limbs, resembles that of the modern cheetah more closely than any other cat, suggesting that extinct cheetah-like forms with similar proportions could have lived in the cold mountain ecosystems of Plio–Pleistocene Europe. Some South African cheetahs hunt in irregular terrain, despite their reputation for preferring open plains.

A large relative of cheetahs, *Miracinonyx inexpectatus,* lived in North America from about 3.2 million years ago until 20,000 years ago. Like the European cheetah, it was big; early *Miracinonyx* are very close anatomically to modern pumas and may even be ancestral to them. However, it was less specialized than the living cheetah, which raises the question of what it might have eaten. Its prey were probably larger versions of today's pronghorn antelope and similar bulky grazers.

The living cats most anatomically similar to the cheetah are the puma and the snow leopard—both fairly chunky animals. The basic structure of their skeletons is similar, but the differences arise from the cheetah's adaptation for super-high-speed pursuit. It departed from the cat pattern to become a more doglike sprinter with only partly retractable claws (for better grip), stiff legs, and a small skull. It has retained the highly flexible catlike spine to make it probably the swiftest land predator ever.

What it does not have is strength, as might be expected from its slim lithe build. Cheetahs are adapted to chase only the very fastest of prey—animals that no other carnivores can catch. But they are easily driven away from their kills by more powerful animals such as even a lone hyena. Their bursts of speed are exhausting, and if several successive kills are wrested from them, they become too fatigued to sustain further chases.

At this point an individual cheetah faces potential disaster, especially a female with cubs, which may soon starve if their mother cannot provide food. When prey becomes scarce, as when local ecosystems change, the whole species is affected, because cheetahs are not adapted to kill larger, more powerful animals. Over an extended period of time—perhaps times of low prey density—populations crash and may go extinct. So cheetahs are enormously vulnerable to ecological disturbances, like the various saber-toothed cats before them.

The presence of humans does not help the situation of the cheetah either, with farming, hunting, and loss of habitat areas, all of which limit the cheetah's options for survival.

LYNX AND HARE

In the Arctic regions of Canada, the snowshoe hare is the preferred prey of the Canadian lynx, whose diet consists of 80 to 90 percent snowshoe hare. Populations of these two mammals mirror each other closely, and when harsh winters cause a reduction in the numbers of hares, the lynx population also drops sharply; conversely, an abundance of hares following a mild year results in an upturn in the number of lynxes. This parallel fluctuation in the numbers of prey and predator can be seen on the graph (left).

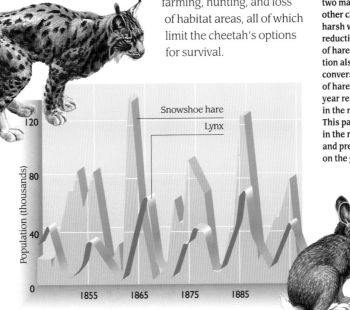

Snowshoe hare

Lynx

Population (thousands)

120

80

40

0

1855   1865   1875   1885

HOLOCENE

INDIVIDUAL groups of organisms have always been at risk in the natural world because of the changes that can occur in the Earth's geosphere, hydrosphere, and atmosphere. Changes in any of these systems can utterly disrupt established organisms, and mass extinction events show this to have happened at various times throughout the Earth's long history. Despite at least five well-documented mass extinctions, and a considerable number of smaller extinction events that have taken place during the past 500 million years, life has recovered eventually.

Extinctions are part of nature, but they are increasingly becoming human-made.

EXTINCT FAUNA OF MAURITIUS

Europeans who arrived on Mauritius in the sixteenth century encountered the dodo, a large flightless bird of the pigeon family, unique to Mauritius. It had no natural predators on the island and so had discarded its ability to fly; in fact, it lacked even the instinct to run away. Dodos were thus easy prey for humans, who hunted them for meat, and for the dogs who accompanied Dutch settlers in the mid-seventeenth century. Cats and rats helped to dispose of the eggs and chicks while men and dogs hunted the adult birds. By 1680 the dodo was extinct. The same fate befell the blue pigeon (pigeon hollandaise), which could fly, but whose eggs were hunted from its treetop nests by the monkeys introduced by colonists. Another loss was the massive domed giant tortoise, weighing up to 100lbs (45kg). Hunted by humans, it also became the prey of imported pigs, which killed young tortoises and dug up eggs from the sand in which they were incubating.

Continued disruption of the global environment by humans may be one step beyond the abilities of other organisms to bounce back. It is extremely unlikely that the consequences of human activities could wipe out microbes, fungi, algae, and other organisms that have been found in environments as extreme and far apart as several miles below the ocean bed and the edges of the stratosphere. But for larger, more conspicuous organisms such as trees, mammals, birds, and reptiles, the outlook is very different. These organisms require large habitable areas of complex micro-environments that are often very delicately balanced. One species of algae has spores that simply recolonize other rock surfaces away from areas of human-generated disturbance. Larger animals cannot do this. Once their habitats have gone for good, so are they. Examples include many types of bird, mammal, reptile, and tree that live in the forested areas of South America, which have been deforested by unrestricted logging.

Humans have had a major part to play in the evolution of animal life for the past 500,000 years or more. Our impact has by no means been confined to the time span since the Industrial Revolution: the dodo, a large flightless dove, became extinct about 100 years before industrialization of the northern hemisphere began, and its home was the Indian Ocean island of Mauritius, far from the scene of the first factories. Like other small islands scattered around the globe, Mauritius had probably been visited by sailors from various regions, but it was uninhabited by humans when Europeans arrived in the early sixteenth century. Forests of ebony covered the mountains, and flocks of giant turtles numbering in the thousands lay on the beaches. The island was dominated by birds, many of which had no predators, and so had no instinct for self-defense. The dodo was just one Mauritian species that vanished.

HOLOCENE

This is the problem facing endangered species with small populations, and the human experts who try to restore the numbers of individuals to viable levels.

As their habitats are interfered with, similar animals may experience contrasting fortunes: for instance, the black-footed ferret of the North American plains did not have access to the new urban ecosystems that have become available to the European stone marten. If the roles of these mustelid carnivores were reversed, the outcomes would probably have been very similar, although the longer legs and tree-dwelling habit of the marten would have given it different advantages.

Other species seem all but impervious to the effects of humans—even to humans' concerted efforts to eliminate them. Rats and cockroaches fall into this category. Unlike the highly specialized cheetah and black-footed ferret, most "pests" are able to retain full populations in spite of environmental disturbances because there is nothing specialist about their lifestyles. To judge from their fossil record, which shows that they have survived at least two mass extinctions, frogs, small lizards, and chelonians (turtles and tortoises) are also highly stable animals that sail through major changes in ecosystems. All of these non-endangered animals are relatively small.

Large animals—like the dodo—make conspicuous targets. Four species of North American bison once lived and grazed from Oregon as far east as Pennsylvania in herds of a million animals or more, grazing on areas more than 1000 square miles (2600 square kilometers). As the plains were settled by Europeans and their descendants, the great open ranges disappeared, and bison were wantonly shot; "Buffalo Bill" Cody personally dispatched 4862 animals in a single year. With the full encouragement of the United States government, which wished to subdue the Native Americans who depended on bison for meat and hides, at least 75 million animals were killed between 1850 and 1880. The Oregon and

THE DISAPPEARANCE of species through habitat loss or reduction is an indirect effect of human interference. Human effects on organisms may also be direct, as in the continued hunting of whales for blubber and associated products, which has reduced many populations to critically low levels. When so few individuals are left, there is insufficient variation in their gene pool (the total genetic material of the population), and fewer opportunities for adaptation to changing circumstances. As the population dwindles, inbreeding reduces the number of healthy offspring, and damaging mutations begin to appear in subsequent generations.

Hunting not only reduces the number of individual animals, it also diminishes the gene pool, limiting the variety and adaptability of a species.

THREAT TO SEALS

(Above) Seal bones in the Arctic. Excessive hunting of seals for their fur has caused a collapse in some populations, while pollution of their habitat is endangering others.

BISON RANGES

(Below) Before the arrival of Europeans, bison grazed from one end of North America to the other; by 1875 they were concentrated in two groups in some of the more remote terrain. Most surviving bison today are in herds of fewer than 500 animals.

1 Mauritius blue pigeon (pigeon hollandaise)
2 Broad-billed parrot
3 Domed Mauritian giant tortoise
4 Dodo
5 Mauritian red rail

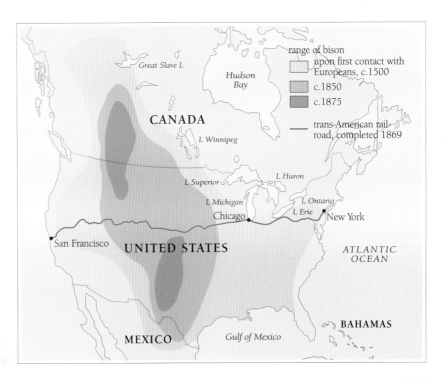

range of bison
upon first contact with Europeans, c.1500
c.1850
c.1875

trans-American railroad, completed 1869

Great Slave L.
Hudson Bay
CANADA
L Winnipeg
L Superior
L Huron
L Michigan
L Ontario
L Erie
Chicago
New York
San Francisco
UNITED STATES
ATLANTIC OCEAN
BAHAMAS
MEXICO
Gulf of Mexico

Pennsylvania species went extinct, but the two remaining American bison were successfully protected and conserved from the beginning of the twentieth century. About 30,000 Great Plains bison (*Bison bison bison*), the smallest species, survive today, while the Wood bison (*Bison bison athabasca*) of the north remains endangered.

Large animals are conspicuous and often beautiful, which helps to bring them sympathetic attention from the public. Less spectacular species are in a more perilous position. They include many species of insects and plants which are becoming scarce as humans devastate their habitats. Because a worm or moss is less visible than a magnificent tiger or redwood tree, the disappearance of the former tends to be discovered too late.

ONE group of animals may yet prove to have a greater effect on global climates than the ice ages in which they evolved: humans. By taking materials from nature for food, shelter, clothing, tools, and other goods, people have been altering their environment for more than a million years. At first, the impact made by small bands of hunter–gatherers was no more than that of most other large animals. It was only in the most recent 10,000 years that technological advances by humans began to have profound effects on the landscape and environment.

Human impact on the environment began with settled farming: the axe, the plow, and livestock.

The first of these innovations was settled farming, which began about 8000 years ago in the Middle East, southeast Asia, China, and Central and South America. Large areas of wild plants were lost as land was plowed up for crops or used to graze domesticated livestock. Entire forests began to be cut down for fuel and timber. Terraces were built on hillsides, and irrigation schemes diverted the courses of rivers in order to water crops. The food surpluses engendered by these technological advances allowed human populations to grow dramatically, increasing the impact on the environment, and eventually sending people in search of ever more land to support them. With European colonization after 1500 AD, new ground came under cultivation throughout the Americas, Australia, and New Zealand. Between 1700 and 1850, farmland around the world doubled from 655 million acres (265 million hectares) to 1.3 billion acres (537 million hectares).

The consequences of these activities are only now becoming fully understood. The area of the Earth that is suitable for growing food is limited: half of the planet's surface is covered by ice, snow, desert, and mountains. Most of the world's population lives on and farms a mere 21 percent of the land, which has come under increasing pressure, particularly in the last 200 years. As much as a third of this land is at risk of becoming non-productive due to overplanting and overgrazing.

Forests have decreased by one-third since the advent of agriculture, and are disappearing faster than any other biome, at a rate of up to 50 million acres (20 million hectares) a year as trees are cleared to support industry as well as farming. Much of this is tropical rainforest,

HARVEST DEITY?

This stone carving (left) from Hungary shows a male, possibly a deity, with a sickle. It dates from the third or fourth millennium BC, when cereal farming was well established in Europe.

THE SPREAD OF AGRICULTURE

(Below left) Farming appears to have arisen independently at different stages in several parts of the world. Climate played a part, with increased rainfall at the end of the last ice age encouraging the growth of wild cereals in the Middle East, where archaeological evidence of farming is especially rich. Mexico, northern Asia, and the Balkans were also sites of very early developments.

HOLOCENE

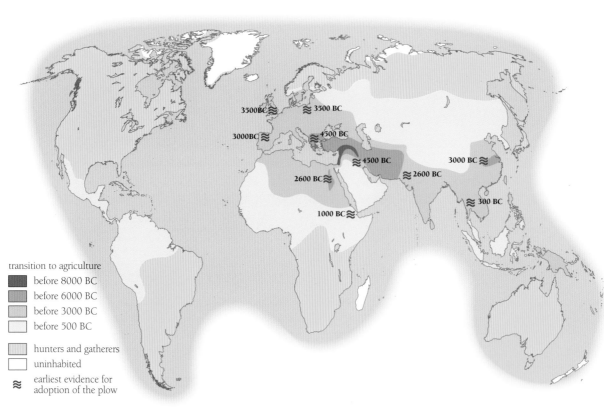

3500BC
3500 BC
3000BC
4500 BC
4500 BC
3000 BC
2600 BC
2600 BC
1000 BC
300 BC

transition to agriculture
- before 8000 BC
- before 6000 BC
- before 3000 BC
- before 500 BC

- hunters and gatherers
- uninhabited
- ≈ earliest evidence for adoption of the plow

(Right) Human activity
has caused the loss of as
much as one-third of the
world's forests in the past
10,000 years. Clearing of
forests causes the loss of
mature trees and associ-
ated plants. Topsoil loses
its binding and the land is
rapidly denuded, a pro-
cess almost impossible to
reverse. Runoff and sedi-
ment clogs rivers, and
land is turned into desert.
Huge areas of the world
are now under this threat.

CUT DOWN
Clearing of forest (below)
contributes to global
warming, as plants trap
carbon dioxide gas that is
otherwise released into
the atmosphere. Defor-
estation is thus a crucial
environmental issue.

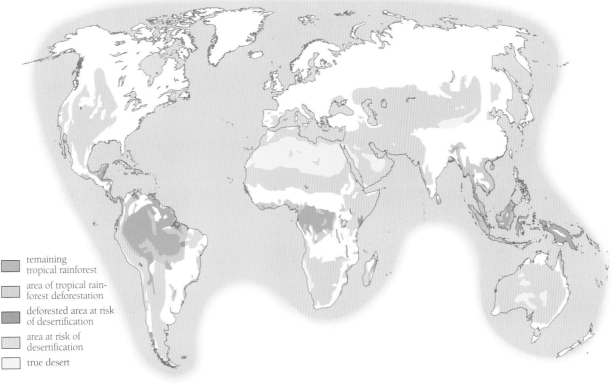

remaining
tropical rainforest

area of tropical rain-
forest deforestation

deforested area at risk
of desertification

area at risk of
desertification

true desert

which supports at least 50 percent of all species of plants
and animals now living. Forests have taken hundreds or
thousands of years to develop to their present complex
maturity; even if an area is replanted when logging is
over, the new forest will take many generations to come
to maturity, and species may be lost in the meantime.
Removal of trees also kills the understory plants, and as
roots that previously bound the top surface of the soils
are removed, water runs off more easily. Intensive
grazing by animals has a similar effect, badly damaging
land by loosening the soil. Eroded topsoil is degraded,
leaving a barren area with almost zero productivity.

Stripped of the vegetation that holds them together,
soils are washed away by rain, and deep gulleys form,
reducing the area that can be cultivated. The eroded soil
silts up rivers, waterways, and dams, interfering with the
water supply; or it blows away, sometimes forming
massive dust and sand storms, as in Saharan Africa and
on the central plains of North America in the 1930s.
Attempts to replace topsoil are not effective and often
make the problem worse; the added topsoil undergoes
further denuding and erosion, while plants find it difficult
to root themselves. Rapidly decaying plant matter is
washed away and clogs nearby streams. In arid and semi-
arid areas, eroded land quickly turns into desert. In the
1990s the United Nations estimated that desertification
affected nearly one-third of the world's land surfaces and
could threaten the livelihoods of 850 million people.
Access to water supplies from major bodies of water is
becoming a sensitive political issue from the arid Middle
East and Africa to parts of Asia and North America.

However much humans and other modern species are
affected by these events, they may be regarded as secondary: even if most species disappeared, new ones would
evolve to take their places. The other great issue of
ecology is how much the planet itself is being damaged.

HOLOCENE

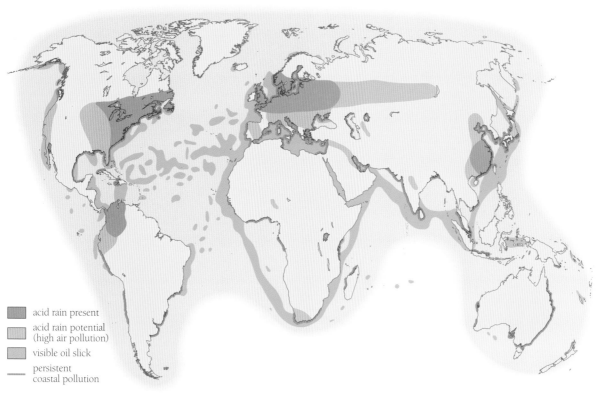

acid rain present

acid rain potential
(high air pollution)

visible oil slick

persistent
coastal pollution

HOLOCENE

THE GREENHOUSE EFFECT

(Right) Energy from the Sun is absorbed by the Earth and radiated back into space. However, mos of the heat is prevented from escaping by "greenhouse gases" such as carbon dioxide, carbon monoxide, and methane in the lower atmosphere. In natural conditions (1), most carbon dioxide—produced through decay, respiration, and chemical weathering—is absorbed by trees in forests and by phytoplankton in the seas. (2) As humans create more and more carbon dioxide through the burning of fossil fuels it accumulates in the atmosphere. There it absorbs the outgoing solar energy, trapping heat so that global temperatures begin to rise.

BEGINNING with largescale industrialization in the nineteenth century, human activities moved beyond overuse of land to the pollution of entire ecosystems.

Pollution now affects all of the biosphere: air, water, and land. No part of the Earth is still beyond reach of contamination.

Not only the land but the whole atmosphere, hydrosphere, and geosphere are at risk. Thousands of tons of hazardous and toxic chemicals are poured into the atmosphere every day, mostly from factories and vehicle emissions. These wastes combine to produce acid rain and smog, and contribute to the depletion of the ozone layer and to global warming.

Acid rain forms from sulfur dioxide, nitrogen oxides, and hydrocarbons. It increase the slight natural acidity of rain, increasing in turn the acidity of the soil or water on which it falls. Toxic elements such as aluminum and cadmium are leached from the soil and absorbed by plants, which lose leaves and then branches; whole forests have died in Eastern Europe and northeastern

DIFFERENT FORMS OF POLLUTION

(Above) There are few places left on Earth that are not affected by the major forms of pollution. Oil slicks or industrial and sewage pollution follow the coast of every continent, while inland, acid rain is the plague of industrial areas, and is spreading as air pollution becomes more prevalent. Land areas not affected by industrialization are often at risk of desertification.

GREAT BARRIER REEF

Australia's Great Barrier Reef (right) is one of the most fragile ecosystems on Earth. Its 400 species of coral support another 5500 species, all of which are vulnerable to even slight changes in the water temperature.

OIL SLICKS AND SEABIRDS

Maritime accidents have a disastrous effect on marine life (left). Seabirds and shore birds are particularly vulnerable to oil slicks, which are a major cause of mortality. Petroleum destroys the natural coating on their feathers, which lose their waterproofing; oil-slicked birds are unable to float. Ingestion of oil is usually fatal to those birds that do not drown.

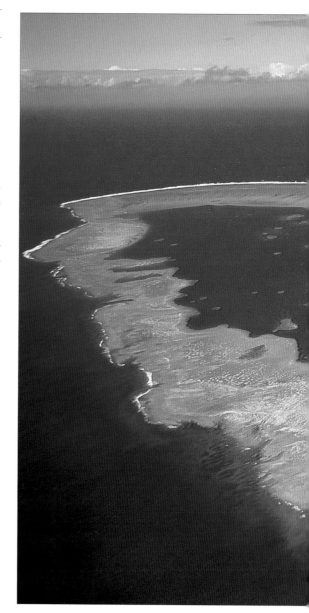

North America. Those plants that survive are left in a weakened condition, unable to tolerate drought, frost, or disease. Acid rain also kills fish and invertebrates, leaving lakes and rivers "dead." Fresh water is a particularly precious resource; only 0.25 percent of all the water on the planet is of use to humans. The rest is too salty.

The ozone layer in the upper atmosphere, 9 to 12mi (15 to 20 km) above the Earth, absorbs harmful ultraviolet radiation from the Sun. Chemicals such as chlorofluorocarbons (CFCs) react with the ozone, causing it to break down. CFCs were extensively used in aerosols and refrigerants until the 1980s, and they will remain in the atmosphere for many decades. Their presence has thinned the ozone layer, as seen in a "hole" over Antarctica in the spring. In 1987 the hole was the size of the United States, and it appears to be increasing, letting in ever more ultraviolet radiation.

Chlorofluorocarbons are among the "greenhouse gases," which also include carbon dioxide and methane. They trap heat from the Earth that would otherwise be radiated back out into space, thus raising the temperature of the planet. Without these gases, it is thought that

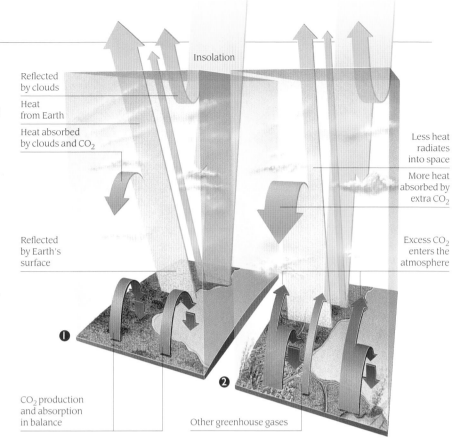

Insolation

Reflected by clouds

Heat from Earth

Heat absorbed by clouds and CO_2

Less heat radiates into space

More heat absorbed by extra CO_2

Reflected by Earth's surface

Excess CO_2 enters the atmosphere

❶

❷

CO_2 production and absorption in balance

Other greenhouse gases

the average temperature around the world would currently be 0.4°F (−18°C). Increased levels of greenhouse gases in the atmosphere over the past hundred years seem instead to have raised the temperature by an average of 1°C (0.5°C). This sounds minor, but most scientists agree that if carbon dioxide levels continue to rise at the present rate, the Earth could warm up by 5.4° to 10°F (3° to 5°C). The last comparable temperature change was during the last ice age —and it took place 10 to 100 times more gradually. This may prevent the onset of the next ice age, or modify it, but consequences are likely to be severe, bringing worse flooding and storms to some areas and drought to others.

The oceans have an immense capacity to soak up and recycle many substances, including carbon dioxide. However, the oceans are being abused on no less a scale than the land and air, receiving forms of waste that are considered too toxic for land disposal: nuclear waste has been dumped at sea since the harnessing of atomic energy in the 1940s. The problem of safe disposal remains one of the great drawbacks of converting to nuclear energy to reduce the burning of fossil fuels on which the industrialized nations are so dependent.

Given the fluctuations that have taken place in the hydrosphere and geosphere during the past 3 billion years, can humans really put the entire planet at risk? If the hydrosphere, atmosphere, and geosphere are considered separately from the biosphere, then the answer is probably "No." Even if all life became extinct, the physical cycles of the planet would continue much as they have for the last four billion years or so, at least until the Sun burns out, another five billion years from now. But it would be supremely ironic if the species that has evolved to this level, priding itself on its "superiority" to other animals by virtue of its great brainpower, turned out to be the one that wipes out the others on which it depends.

MODERN-DAY EXTINCTIONS

PHOTO OPPORTUNITY

The plight of the endangered blue whale, and other large mammals, has captured the public imagination. Yet even more at risk of extinction are many smaller, less celebrated creatures, such as rats and bats, fish, and numerous invertebrates.

EXTINCTIONS are part of the normal meta-cycle of life. The fossil record is full of species that appeared and then vanished—about 95 percent of all that have ever existed. Factors such as body size and level of specialization of diet are important in directing which groups become extinct under "normal" circumstances. This probably accounts for a steady rate of extinction of one species per year over the past 3.5 billion years. In addition there have been mass extinctions due to catastrophic events such as meteorite impacts, or to periodic global climate changes. The latter have always been particularly devastating to the richly diverse plants and animals of tropical regions.

Modern extinctions are almost entirely the direct or indirect consequences of human action. Some species are hunted; others die out as their habitats are polluted, or cleared for farming or development. Yet others cling to existence in dwindling populations too small to be viable for breeding, especially for large animals that produce only one offspring at a time, at infrequent intervals.

According to the most conservative estimate of the distinguished American biologist Edward O. Wilson, in the 1990s species were becoming extinct at a rate of three per hour, or 27,000 per year. Within 30 years this figure may rise to *several hundred species per day*, with consequences that are truly terrible to contemplate. Every plant that goes extinct, for instance, may take with it as many as 30 insects and animals that depend on it for food.

No one is sure how many species currently occupy the planet, but the figure is believed to be between five and ten million. In the past 400 years, 611 species of animals alone are known to have disappeared, representing 1.8 percent of all mammals and one percent of birds at the absolute minimum, but possibly more. Birds and mollusks account for most animal extinctions, with mammals in third place. Of mammals, rodents and bats are disappearing fastest and have the highest number of endangered species. Primates also figure prominently among animals at risk, although there have not yet been any proven cases of extinction among them.

The Earth itself is resilient; life in one form or another will probably go on. After the latest mass extinctions, new species of rapid-spreading opportunists will arise to fill the empty niches. Weeds are a familiar example. So are starlings, a successful but unpopular bird; also cyanobacteria, which flourish in polluted water, lowering its oxygen content, and smelling foul. If humans survive the devastation we have wreaked, we may not enjoy sharing the planet with such species while waiting for wild flowers, dolphins, and pandas to re-evolve. But the human population itself will most likely be severely depleted as well, as our own habitats turn to desert or become flooded by the sea as polar ice melts, and food becomes scarce in many parts of the world.

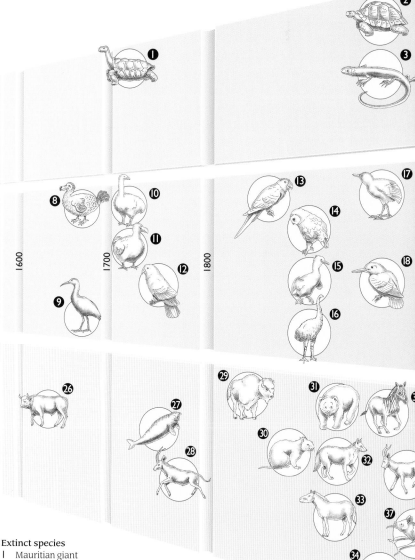

1600   1700   1800

Extinct species

1 Mauritian giant tortoise
2 Charles Island tortoise
3 Réunion skink
4 Palestinian painted frog
5 Abingdon Island tortoise
6 New Zealand grayling
7 Round Island boa
8 Dodo
9 Mauritian red rail
10 Elephant bird
11 Réunion solitaire
12 Guadeloupe amazon
13 Green and yellow macaw
14 Rodriguez little owl
15 Steller's spectacled cormorant
16 Giant moa
17 Jamaican wood rail
18 Ryukyu kingfisher
19 Passenger pigeon
20 Carolina parakeet
21 Lord Howe Island white eye
22 Pink-headed duck
23 Hawaiian o-o
24 Madagascar serpent eagle
25 Kauai nukupuu
26 Aurochs
27 Steller's sea cow
28 Blue buck
29 Eastern bison
30 Hispaniolan huita
31 Atlas bear
32 Falklands wolf
33 Tarpan
34 Sea mink
35 Quagga
36 Portuguese ibex
37 Jamaican long-tongued bat
38 Arizona jaguar
39 Newfoundland white wolf
40 Barbary lion
41 Greater rabbit-bandicoot
42 Syrian onager
43 Schomburgk's deer
44 Thylacine
45 Bali tiger
46 Toolache wallaby
47 Caribbean monk seal
48 Mexican silver grizzly bear

HOLOCENE

Fish, amphibians, reptiles

Birds

Mammals

2000

Critically-endangered species

49 Totoaba
50 Common sturgeon
51 Chinese alligator
52 Tinkling frog
53 Grand Cayman blue rock iguana
54 Santa Cruz long-toed salamander
55 Antiguan racer
56 Milos viper
57 Golden toad
58 Mallorcan midwife toad
59 Hawksbill turtle
60 Western swamp turtle
61 Californian condor
62 Whooping crane
63 Eskimo curlew
64 Juan Fernandez firecrown
65 Northern bald ibis
66 Spix's macaw
67 Seychelles magpie-robin
68 Spotted owl
69 Pink pigeon
70 Night parrot
71 Gurneys pitta
72 Black spoonbill
73 Bali starling
74 Black stilt
75 Zapata wren
76 African wild ass
77 Baluchistan bear
78 Père David's deer
79 Rodriguez flying fox
80 Gray whale (Asian stock)
81 Kouprey
82 Broad-nosed gentle lemur
83 Sumatran orang-utan
84 Florida panther
85 Visayan warty pig
86 Black rhino
87 Mediterranean monk seal
88 Golden crowned sifaka
89 Ethiopian wolf
90 Hairy nosed wombat

RECORDING LOSS

The fossil record shows that the average lifespan of a species is 5 to10 million years. With 5 to 10 million living species, the extinction rate should be one a year. While it is hard to know exactly the current rate of loss it is clear that it is well in excess of "normal". Over the last 400 years there were 611 documented extinctions, but this record excluded many creatures including most invertebrates—which account for 95 percent of all animals. Today, there are over 5000 threatened species listed, but only a very small proportion of recognized species have been evaluated.

HOLOCENE

GLOSSARY

A

ABYSSAL PLAIN The floor of the ocean where it forms a broad plain between 2 and 4mi (3 and 6km) below sea level.

ACADIAN OROGENY The mountain-building event that formed the northern Appalachians in the Devonian period. In Europe it is known as the CALEDONIAN OROGENY.

ACANTHODIAN A jawed fish of the Silurian–Carboniferous with a short, blunt head and prominent spines in front of each fin.

ACCRETION In PLATE TECTONICS, the building up of a CONTINENT when volcanic ISLAND ARCS become fused to the edge of a landmass.

ACCRETIONARY BELT (wedge) A part of a CONTINENT formed by ACCRETION of ISLAND ARCS, TERRANES, and fragments of emergent OCEANIC CRUST.

ACHEULEAN An early Pleistocene culture of stone tool-making consisting of roughly-worked stone blades. It has been attributed to early *Homo erectus* or *Homo habilis.*

ACID RAIN Rain that has become acidic due to the presence of dissolved substances such as sulfur dioxide. This may be caused by volcanic eruptions or, in modern times, industrial pollution.

ACRITARCH A planktonic microalga, usually with an ornamented envelope, that existed from the Proterozoic through the early Cenozoic. Most acritarchs were probably related to DINOFLAGELLATES.

ACTINOPTERYGIAN A member of the subclass of fish that have fins with radiating supports. Most living fish are actinopterygians.

ADAPIFORME A member of a group of primitive lemurs that lived in early Tertiary times.

ADAPTATION In EVOLUTION, the changing of an organism's structure or behavior to allow it to live a particular lifestyle in a particular environment, such as ducks' webbed feet.

AGNATHAN *See* JAWLESS FISH.

AGRICULTURAL REVOLUTION The change in farming practices in Europe in the late eighteenth and early nineteenth centuries, when scientific practices began to be applied to the cultivation of large areas and food production increased dramatically.

ALBEDO The amount of light reflected from a body, particularly from different areas of Earth or from the moon or a planet.

ALGA (plural algae) The most primitive form of plant, consisting of a single cell or a mass of cells but without a plumbing system. Seaweeds are an example of algae.

ALLEGHENIAN OROGENY A continuation of the ACADIAN OROGENY that occurred in the Late Paleozoic as three continents encroached on LAURENTIA, forming the ancient Appalachians. The European extension of this event is known as the Hercynian orogeny.

ALLOCHTHONOUS TERRANE *See* EXOTIC TERRANE.

ALLUVIAL FAN A fan-shaped sheet of eroded material washed down from an upland area and deposited over a flat plain.

ALPINE OROGENY The Tertiary collision of Europe with Africa, closing the Tethys Ocean between them and raising the Alps.

ALTAY SAYAN FOLD BELT The mountain system in Central Asia which was pushed up during the accretion of southern Siberia and Mongolia to northern Siberia.

ALULA A group of feathers on a bird's wing, in the position of the thumb, that helps maneuverability in flight.

AMINO ACID An organic compound based on NH_2 and COOR that forms the basis of PROTEIN and hence the basis of all living things. There are about twenty different types of amino acid.

AMMONITE One of a group of AMMONOID CEPHALOPODS common in the Mesozoic, having mostly coiled shells and extremely complex suture lines. Their range and rapid evolution make them ideal INDEX FOSSILS.

AMMONOID The extinct group of CEPHALOPODS to which AMMONITES belonged, along with goniatites and ceratites.

AMPHIBIAN The most primitive form of TETRAPOD, which passes a larval stage in the water and a mature stage usually on land. Frogs and newts are examples.

ANAPSID A member of one of the major subclasses of REPTILES, defined by the absence of holes in the skull behind the eye socket. Anapsids are regarded as the most primitive of the reptile groups; turtles are examples.

ANASPID A member of an order of JAWLESS FISH of the Silurian and Devonian which lacked the heavy head armor of many early fish.

ANDESITE A fine-grained IGNEOUS rock, gray in color, consisting predominantly of oligoclase or FELDSPAR. It is particularly abundant in the Andes, for which it was named.

ANGARALAND The continent formed by the collision of the individual islands of Kazakhstania and Siberia during the Permian. It in turn became attached to LAURASIA as the Uralian Seaway closed.

ANGIOSPERM The technical term for a flowering plant—one that bears its seeds in a box such as a pod or a fruit.

ANGULAR UNCONFORMITY A type of stratigraphic UNCONFORMITY in which overlying horizontal rock strata are separated from the older, tilted, eroded beds below.

ANKYLOSAUR One of a group of four-footed ORNITHISCHIAN dinosaurs characterized by an armored covering on the back, and either a bony club on the tail or an array of protective spikes at each side.

ANNELID WORM An INVERTEBRATE with an elongated body which is a muscle sack subdivided into segments with a recurrent set of organs. Annelids existed from the Cambrian.

ANOMALOCARIDID A Cambrian marine INVERTEBRATE predator. It had a huge head on which there was a pair of compound eyes on its upper surface and a rounded mouth with two spiny appendages below.

ANOXIA The state in which water contains less than 0.003fl.oz. of oxygen per 0.22 gallon (0.1ml oxygen per liter), the threshold below which animal life diminishes significantly.

ANTHROPOID One of a group of highly derived PRIMATES that appeared in the Paleogene period. It includes apes, monkeys, and all HOMINOIDS.

ANTLER OROGENY The mountain-building event that produced a contemporary range from Nevada to Alberta, North America, in Late Devonian and Early Carboniferous times.

APLACOPHORAN A worm-like marine mollusk lacking a shell and foot. Only modern species are known.

APPALACHIAN OROGENY The prolonged Late Paleozoic mountain-building event produced by the long collision between Laurentia (North America), Baltica (northern Europe), and Gondwana. It encompassed the TACONIAN, ACADIAN, and ALLEGHENIAN orogenies, which built the Appalachians.

ARCHAEOCETE A member of a family of early whales from the Early Tertiary. They had teeth of different shapes and sizes, and most had long serpentine bodies.

ARCHAEOCYATH A cuplike sessile marine animal of the Cambrian. It was a possible relative of DEMOSPONGES.

ARCHAEOPTERYX The earliest known bird, found in Upper Jurassic rocks of Germany, which retained many of the anatomical features of its ancestors, the DINOSAURS.

ARCHEAN The first eon of geological time, comprising about 45 percent of Earth's history, from 4.55 billion to 2.5 billion years ago.

ARCHOSAUR A member of a group of DIAPSID REPTILES encompassing the crocodiles, the DINOSAURS, the pterosaurs, and the birds.

ARTHROPOD An INVERTEBRATE animal with an articulated external skeleton covering a segmented body which bears serial pairs of articulated appendages. Arthropods first appeared in the Cambrian and have survived to the present; they include spiders, mites, CRUSTACEANS, centipedes, and insects.

ARTIODACTYL An even-toed, cloven-hoofed, grazing or browsing UNGULATE, such as pigs, deer, and cattle. *See also* PERISSODACTYL.

ASTEROID A minor planet in orbit around the Sun. Most lie between the orbits of Mars and Jupiter and vary in diameter from about 10mi (16km) to over 500mi (800km).

ASTHENOSPHERE The mobile portion of the Earth's inner layers, about 30 to155mi (50 to 250km) below the surface, on which the TECTONIC PLATES move.

ATMOSPHERE The envelope of gases surrounding the Earth, keeping it warm enough for life to survive, and filtering out harmful ultraviolet rays from the Sun. The most prevalent gas is nitrogen (78 percent), and the most important for life are oxygen (21 percent) and carbon dioxide (0.03 percent).

AUSTRALOPITHECINE A member of a group of Plio–Pleistocene HOMINIDS with anatomy intermediate between apes and humans.

AUTOTROPH Any organism that is a producer of food: that is, plants.

AVALONIA A continent that coalesced in the early Paleozoic and was amalgamated into LAURENTIA and BALTICA in the Late Paleozoic. Its components included modern eastern Newfoundland, the Avalon Peninsula and Nova Scotia (in North America), southern Ireland, England, Wales, and some fragments of continental Europe: parts of northern France, Belgium, and northern Germany.

B

BACK ARC BASIN The sea behind the arc of islands formed by volcanic action as one of the Earth's plates overrides another.

BACK REEF The landward side of a reef, including the area behind the reef crest and shelf lagoon.

BACTERIOPLANKTON PLANKTONIC bacteria.

BACTERIUM One of a group of unicellular prokaryotic microorganisms related to the fungi. Bacteria appeared between 3.5 and 3.8 billion years ago and are one of the most successful life forms on Earth.

BALEEN A horned comb-like structure in the mouths of the toothless whales, used for filtering small animals from the seawater.

BALTICA A continent of the Paleozoic and Mesozoic which included eastern and northern Europe surrounding the Baltic Sea.

BANDED IRONSTONE A Precambrian rock of alternating iron-rich and iron-poor layers.

BARNACLE A sessile filter-feeding marine CRUSTACEAN with a multi-plated calcareous shell. Barnacles originated in the Silurian and became widespread in the Triassic.

BASALT A dark IGNEOUS rock in OCEANIC CRUST, consisting mostly of PLAGIOCLASE, pyroxene, and glassy substances.

BASE PAIR A pair of nucleotide bases, joined by hydrogen bonds, that holds together the double strands of DNA and some RNA. Their units are pyrimidines (thymine, cytosine, or uracil) and purines (adenine or guanine), which are components of nucleic acids.

BASIN A low-lying geographic area that tends to collect sediments from higher land around it, thus building up rock sequences.

BATHOLITH A large, typically irregularly shaped body of intrusive IGNEOUS rocks, often granitic, with an exposed surface area of more than 40mi^2 (100km^2).

BED A layer of SEDIMENTARY ROCK that is distinct from the layers above and below.

BEDDING PLANE The surface that separates one bed of SEDIMENTARY ROCK from the next.

BELEMNITE One of a group of Mesozoic squid-like CEPHALOPODS.

BENIOFF ZONE The steeply inclined zone of seismic activity that extends down from an OCEAN TRENCH toward the ASTHENOSPHERE. Such zones mark the path of a TECTONIC PLATE being subducted at a destructive plate margin. The foci of earthquakes become deeper toward the non-subducting plate, reaching more than 370mi (600km) deep.

BENTHIC A bottom-dwelling aquatic organism.

BERING LAND-BRIDGE The span of land that is exposed periodically across the Bering Strait, connecting North America and Asia.

BIG BANG The theory that proposes the origin of the entire Universe (including space, matter, energy, time, and the laws of physics) about 15 billion years ago in the explosion of an extremely hot, dense body. The debris of the explosion moved radially away from the point of origin, cooled, and eventually formed galaxies and stars.

BIODIVERSITY A measure of the variety of the Earth's species, of the genetic differences within species, and of the ecosystems that support those species.

BIOGENESIS The principle that living things can only evolve out of living things like themselves rather than being created spontaneously or transformed from other things.

BIOGEOGRAPHIC PROVINCE An area with a distinct suite of animals and plants produced by geographical barriers that prevent the inhabitants from mixing with neighboring floras and faunas. Provinces may be recognized by the distribution of unusual life forms such as marsupials, which exist exclusively in Australia in modern times.

BIOMASS The total mass of living organisms in a defined area.

BIOME A broad community of plants and animals shaped by common patterns of vegetation and climate. Grassland, desert, tundra, and rainforest are examples.

BIOSPHERE The part of the Earth that supports life, beginning with the lower ATMOSPHERE and extending through the surface (land and water) to the upper fraction of CRUST.

BIOSTRATIGRAPHY The division of rocks into zones based on their fossil contents.

BIPEDALISM The ability to walk on two legs.

BIVALVE An aquatic MOLLUSK, covered with a calcareous shell of two valves and lacking a distinct head, dating from the Cambrian.

BLACK SMOKER A jet of hot mineralized water rising from vents in the ocean floor where RIFTS form. The color is from dissolved sulfides of iron, zinc, manganese, and copper.

BOROPHAGINE Hyena–dog: a member of a group that evolved from the Tertiary VULPAVINE carnivores and split from the canids (true dogs) in the late Paleogene.

BRACHIOPOD A solitary bivalved sessile marine animal that catches food particles with tentacles on a loop-shaped organ (lophophore). Brachiopods evolved in the early Cambrian before the true MOLLUSKS.

BRECCIA A coarse-grained SEDIMENTARY ROCK made up of angular fragments.

BRONTOTHERE A member of a group of rhinoceros-like ungulates, some of them very large, that existed in early Tertiary times.

BRYOPHYTE A simple land plant with leaves and a stem but no VASCULAR system, such as mosses and liverworts.

BRYOZOAN A colonial sessile marine INVERTEBRATE that traps food particles with tentacles occurring on a lophophore.

BURGESS SHALE The most famous of Cambrian LAGERSTATTEN, located in the Middle Cambrian rocks of western Canada.

C

CALCITE A mineral consisting of calcium CARBONATE.

CALCRETE A layer of LIMESTONE formed on or below the surface of soil caused by the evaporation of CALCITE-rich ground water. It is sometimes called caliche or kunkar.

CALEDONIAN OROGENY The mountain-building event that formed the Scottish Highlands and the Norwegian mountains during the Devonian period, when Baltica and Laurentia collided.

CAMBRIAN EXPLOSION The phenomenal radiation of marine animals during the Cambrian. The Cambrian–Precambrian boundary marked the appearance of nearly all animal phyla, and also a unique array of extinct creatures that cannot be assigned to any recognized phylum.

CARBON CYCLE The sequence of chemical reactions by which carbon circulates through the ecosystem. Carbon is a major component of LIMESTONE, often deposited as the shells of living things. Carbon from carbon dioxide is taken up by plants during PHOTOSYNTHESIS, producing carbohydrates and releasing oxygen into the atmosphere. The carbohydrates are used directly by plants—or by animals that eat plants—in respiration, releasing carbon dioxide back into the atmosphere.

CARBONATE A salt of carbonic acid. Carbonates are common in minerals, and are the principal constituents of SEDIMENTARY ROCKS such as LIMESTONE. The most widespread carbonate minerals are CALCITE, aragonite, and dolomite.

CARBONATE COMPENSATION DEPTH The ocean depth at which the rate of CARBONATE precipitation equals that of dissolution.

CARNASSIAL SHEAR The scissor action of specialized, blade-like molar or premolar teeth in CARNIVORES such as cats and dogs, an evolutionary development that increased their efficiency in cutting flesh.

CARNIVORE Generally, an animal that eats flesh; technically, the term applies only to

MAMMALS of the order Carnivora, encompassing cats, dogs, weasels, bears, and seals.

CARPEL The female reproductive organ of flowering plants from which the fruits and seeds grow.

CATARRHINE A member of the Old World group of broad-nosed monkeys from which humans descended. *See also* PLATYRRHINE (New World monkeys).

CENOZOIC The most recent era of geological time, which began 65 million years ago. It encompasses the Tertiary and Quaternary periods and includes the present day.

CEPHALOCHORDATE A lancelet: a small, scaleless, fishlike, primitive CHORDATE with a NOTOCHORD and a nerve cord but no brain. Cephalochordates arose in the Cambrian.

CEPHALON The head portion of a TRILOBITE.

CEPHALOPOD An advanced marine MOLLUSK, with a large brain and eyes, that evolved in the Cambrian. The foot is developed into a jet-propulsion system and tentacles.

CERATOPSIAN One of group of four-footed ORNITHISCHIAN dinosaurs characterized by the presence of an armored shield and an arrangement of horns on the head.

CERCOPITHECOID A member of a family of primitive CATARRHINE monkeys dating from late Tertiary times.

CETACEAN A member of the mammalian order Cetacea, encompassing the whales, dolphins, and porpoises.

CHAETETID A DEMOSPONGE, possessing a rigid calcareous skeleton, that first appeared in the Ordovician.

CHALK A pure form of LIMESTONE, formed from the deposits of microscopic shelly animals.

CHANCELLORIID A sessile COELOSCLERITO-PHORAN of the Early Paleozoic whose skeleton consisted of rosette-shaped spiny hollow sclerites surrounding a sack-like body.

CHELICERATE An ARTHROPOD such as a water scorpion, spider, mite, or horseshoe crab. Chelicerates appeared in the Ordovician and have bodies subdivided into a head end with six pairs of appendages, the first pair of which are grasping jawlike chelicerae, and a tail portion.

CHEMOSYNTHESIS A process of organic matter production with the use of chemical redox reactions as an energy source. Bacteria are the principal chemotrophs.

CHERT A rock formed of non-crystalline silica.

CHITIN An ORGANIC substance that forms insects' shells and human fingernails.

CHITINOZOAN A problematic Ordovician–Carboniferous PLANKTONIC microorganism with a chitinous casing which is a chain of retortlike individuals; possibly a marine animal egg.

CHITON *See* POLYPLACOPHORAN.

CHLOROFLUOROCARBON A group of synthetic, nontoxic, inert gases containing chlorine, fluorine, carbon, and sometimes hydrogen, used in refrigerants, which accu-mulate in the upper atmosphere and break down ozone.

CHLOROPHYLL The green pigment in plants. Its function is to produce carbohydrates for use as food from carbon dioxide and water.

CHLOROPLAST The structure in a plant cell that contains CHLOROPHYLL.

CHONDRICHTHYAN A jawed fish, first known from the Silurian, whose skeleton consists entirely of cartilage; a shark is an example.

CHORDATE A DEUTEROSTOME with an anterior nerve cord, a cartilaginous rod (notochord) which is replaced by the vertebral column in higher chordates, and gill slits in the throat. Chordates have existed since the Cambrian.

CHROMOSOME A thread-shaped string of DNA that exists in the cells of living things and contains the GENES.

CIRCUMPOLAR CURRENT An ocean current that flows around one of the Earth's poles. In the present there is an important circumpolar current around Antarctica.

CIRQUE (also corrie, cwm) An armchair-like hollow in a hillside that was once the source of a GLACIER and has been widened and deepened by the weight of the ice.

CLADISTICS A method of classification that assigns organisms to taxonomic groups by assessing the extent to which they share characteristics. *See* TAXONOMY.

CLADOGRAM A chart that shows the evolutionary relationships of organisms or groups of organisms by comparing the numbers of features they have in common.

CLASTIC A rock made up of fragments of other rocks or their minerals (such as quartz).

CLASTIC WEDGE A large deposit of clastic sediments produced by uplift nearby. The thicker end of the wedge, and its sediments, occurs closer to the source, and the thinner end further away.

CLAY A very fine-grained SEDIMENTARY ROCK, usually with plastic properties.

CLUBMOSS A primitive VASCULAR PLANT related to the ferns. Nowadays they are small and insignificant, but in late Paleozoic times they grew as trees 300ft (100m) tall.

CNIDARIAN A primitive jellylike aquatic INVER-TEBRATE with stinging cells (nematocysts) on their tentacles. Cnidarians have existed since the latest Proterozoic and include hydras, corals, sea anemones, and jellyfish.

COAL An organic SEDIMENTARY METAMORPHIC rock consisting largely (more than 50 percent) of the carbon remains of plant material. This must accumulate under water or be rapidly buried to prevent oxidation and decay. The depth at which the coal is buried in rock strata, and the resulting pressure, produces either soft, low-grade coal (peat, lignite) or hard, high-grade (anthracite).

COCOS PLATE A small tectonic plate in the eastern Pacific Ocean, bounded by the Galapagos Ridge, the East Pacific Rise, and the landmass of Central America.

COELENTERATE *See* CNIDARIAN.

COELOSCLERITOPHORAN A Cambrian marine animal whose body was covered with hollow scalelike or spinelike SCLERITES. It is a possible ancestor of ANNELIDS, MOLLUSKS, and BRACHIOPODS.

COLD SEEP An area of seafloor on which cool mineral waters seep out through pores and crevices in the rocks.

COMET A planetary body consisting mostly of water, ice, and rock fragments. When its orbit brings it close to the Sun, the water from the ice boils away and forms a tail.

COMMUNITY An entity of interrelated organisms inhabiting a limited area.

CONDYLARTH One of an order of herbivorous PLACENTAL mammals that constituted the majority of MAMMALS in the early Tertiary.

CONGLOMERATE A SEDIMENTARY ROCK that consists of rounded lumps of pre-existing rock cemented together. It is essentially a fossilized shingle beach.

CONIFER A GYMNOSPERM tree, such as fir and pine, that reproduces by means of a cone.

CONODONT A primitive Paleozoic–Mesozoic eel-like swimming marine vertebrate with multiple phosphatized conical teeth.

CONSUMER An animal that eats PRODUCERS or other consumers.

CONTINENT A body of relatively buoyant terrestrial CRUST. The Earth's continents lie on average 2.8mi (4.6km) above the ocean floor and range in thickness between 12 and 37mi (20 and 60km). The oldest continental rocks found so far are approximately 3.8 billion years old. At the heart of each continent are one or more masses of ancient rock called CRATONS or shields, surrounded by successively younger MOBILE BELTS of FOLD mountains. The edges of continents may be flooded to form CONTINENTAL SHELVES.

CONTINENTAL DRIFT A theory, generally attributed to the German meteorologist Alfred Wegener, that postulates the early existence of a single supercontinent that eventually broke up, beginning to drift apart about 200 million years ago—and its components, the continents, are still drifting. Modern research has established that this is the result of SEAFLOOR SPREADING, driven by CONVECTION within the Earth's MANTLE.

CONTINENTAL SHELF A gently shelving, submerged part of a continent's margin that extends from the coastline to the top of the continental slope. Most sedimentation occurs on this part of the ocean floor.

CONULARIID A marine sessile animal of the latest Proterozoic that had a four-sided, elongate pyramidal mineralized skeleton and a four-lobed folded lid.

CONVECTION The movement of a fluid by heat. Hot fluid is less dense than cold, and so it rises; cool fluid descends. Convection currents drive the wind systems of the world as well as PLATE TECTONICS.

CONVERGENT EVOLUTION The phenomenon

by which animals with no recent common ancestry evolve similar shapes or habits through adaptation, enabling them to live the same lifestyle in a similar environment. Ichthyosaurs (reptiles), sharks (fish), and dolphins (mammals) have developed the same shape through convergent evolution, although they are unrelated.

CONVERGENT PLATE MARGIN A region of the LITHOSPHERE in which LITHOSPHERIC PLATES are pushed together and crustal surface area is lost. It may be caused by SUBDUCTION, in which lithosphere is consumed into the MANTLE, or by crustal shortening or thickening, in which slices of lithosphere are stacked upon each other as thrust slices.

COPE'S LAW The principle of EVOLUTION articulated by the American paleontologist Edward Drinker Cope (1840–97), which states that over the course of time all animals tend to evolve larger body sizes.

CORAL Any of a group of marine INVERTEBRATES of the class Anthozoa in the phylum CNIDARIA, and a few species of the class Hydrozoa (hydroids). Corals secrete a skeleton of calcium CARBONATE extracted from water. They occur in warm seas at moderate depths with adequate light. Corals live in a symbiotic (mutually beneficial) relationship with algae, in which the algae obtain carbon dioxide from the coral and the coral receives nutrients from the algae. Most corals form large colonies, although there are some species that live singly, as did early corals, which appeared in the Cambrian. Their accumulated skeletons make up REEFS and atolls.

CORDILLERA A series of parallel FOLD mountain ranges.

CORE The innermost portion of the Earth's surface below the MANTLE, more than 1800mi (2900km) deep, which is believed to be composed mostly of iron, with a solid center surrounded by a molten layer.

CRANIATE Another name for a VERTEBRATE. The term includes animals such as hagfishes which do not have vertebrae but do have the group's specialized skull features.

CRATON A mass of ancient METAMORPHIC ROCK at the center of a continent which is so distorted and compacted that it cannot be deformed further. It is the stable heart of the continent.

CREATIONISM A theory that claims that the world was created by a supreme being not more than 6000 years ago, as stated in the Bible, and that species have individual origins and are unchanging. Developed in response to Darwin's theory of EVOLUTION, it is not considered factual by most scientists.

CREODONT A member of the Creodonta, an order of large carnivorous MAMMALS of the early Tertiary and a sister group to the still-living Carnivora. There were two main lineages, the OXYENIDS and the HYENODONTS.

CRINOID (sea lily) A member of the ECHINO-DERM group, related to starfish, and usually anchored to the sea bed by a stalk.

CROSS-BEDDING (current bedding) Inclined planes in SEDIMENTARY ROCKS caused by strong currents of water or wind during deposition. For example, in a typical delta, where a flowing river drops its sediment load on reaching the deeper water of the sea, there are more or less horizontal or very gently shelving topset beds; inclined foreset beds (the delta front); and gently sloping bottomset beds which meet the flat seafloor in front of the delta. Current flows in the direction of the downward-sloping strata. A similar pattern (DUNE BEDDING) develops where wind forms sand dunes in a desert.

CRUST The outermost portion of the Earth's lithosphere, above the MANTLE. There are two forms: dense OCEANIC or MAFIC crust, and lighter continental or FELSIC crust.

CRUSTACEAN An aquatic gill-breathing member of the ARTHROPOD class Crustacea. Crustaceans evolved in the Cambrian and include crabs, lobsters, shrimps, woodlice, and barnacles. The segmented body usually has a distinct head, thorax, and abdomen, and is protected by an external skeleton made of protein and chitin hardened with calcium carbonate.

CTENOPHORE A radially symmetrical marine invertebrate possessing paddlelike comb plates (giving the common name comb jelly), which propel the animal forward. Ctenophores appeared in the Cambrian.

CUTICLE The hard noncellular protective surface layer of many INVERTEBRATES such as insects. it acts as an EXOSKELETON to which muscles are attached, reduces water loss, and provides defensive functions.

CUVETTE An area of inland drainage between mountains in which beds of SEDIMENTARY ROCK accumulate.

CYANOBACTERIA (blue-green algae) Primitive single-celled organisms structurally similar to bacteria. They are sometimes joined in colonies or filaments. Cyanobacteria are among the oldest known living things, more than 3.5 billion years old, and belong to the kingdom Monera. Their development of PHOTOSYNTHESIS changed the Earth by contributing more oxygen to the atmosphere, allowing advanced life forms to develop. Cyanobacteria are widely distributed in aquatic habitats, on the damp surfaces of rocks and trees, and in the soil. The name derives from the coloration, which is produced by the pigments CHLOROPHYLL and phycocyanin.

CYCADS A diverse group of seed-bearing plants common during the Mesozoic, resembling palms in appearance.

CYCLOTHEM A sequence of beds of SEDIMENTARY ROCK that show that they were deposited in cycles: for instance, LIMESTONE formed in the sea, followed by SANDSTONE deposited as a river encroaches, followed by COAL as plants grew on the river bank, followed by limestone as the sea encroached again.

CYNODONT One of a suborder of flesh-eating MAMMAL-like REPTILES with doglike teeth.

D

DARWINISM A popular name for the theory of EVOLUTION proposed by the British natural scientist Charles Darwin (1809–82). His main argument, now known as the theory of NATURAL SELECTION, concerned the variation that exists between members of a sexually reproducing population. According to Darwin, those members with variations better fitted to the environment would be more likely to survive and breed (survival of the fittest), subsequently passing on these characteristics to their offspring. Over time, the genetic makeup of the population would change; and, given enough time, a new SPECIES would form. Existing species would thus have originated by evolution from older species. See also CREATIONISM.

DEGASSING The process by which gas escapes from a body or a substance. Early in the Archean the Earth was heated to liquefaction and lost a large volume of gases into space.

DEMOSPONGE A class of sponges in existence since the Cambrian. The skeleton is built from spongin (a flexible protein similar to keratin), siliceous spicules, or rigid calcium carbonate, or a combination of these.

DESERTIFICATION The creation of deserts by climate change or by artificial processes including overgrazing; the destruction of forest belts; exhaustion of the soil by intensive cultivation with or without the use of fertilizers; and salinization of soils due to mismanaged irrigation.

DETRITUS FEEDER A consumer that ingests sediment to feed on organic matter; predominantly bacteria.

DEUTEROSTOME ("last mouth") An animal whose embryonic mouth becomes the anus as another mouth develops in the adult. ECHINODERMS, HEMICHORDATES, and CHORDATES are all deuterostomes.

DIAGENESIS The formation of SEDIMENTARY ROCKS from buried sediment at low temperatures. Two processes are involved: first, the particles of the sediment are compacted, and then they are cemented together by minerals.

DIAPIR A dome-shaped rock structure formed by a layer of rock being squeezed up through the overlying beds. This only occurs when that layer consists of a rock that is plastic under pressure, such as salt.

DIAPSID A member of one of the major subclasses of REPTILES. The diapsids are defined by the presence of two holes in the skull behind the eye socket. The lizards and snakes and the ARCHOSAURS are diapsids.

DICYNODONT A member of a suborder of MAMMAL-like REPTILES with a pair of prominent canine teeth.

DIFFERENTIATION (biology) The process by which cells in developing tissues and organs become increasingly different and specialized, giving rise to more complex structures that have particular functions.

DIFFERENTIATION (geology) The formation of distinct types of IGNEOUS ROCK from the same magma. Minerals crystallize at different temperatures and pressures and some accumulate before others, resulting in rocks of different compositions. The primary differentiation of all the planets, from masses of homogenous molten rock to layered planets with cores, occurred in a similar way.

DIKE A small sheetlike intrusion of IGNEOUS ROCK that cuts through pre-existing strata, formed by molten material pushing its way through a crack and solidifying there.

DINOCERATE A member of an order of rhinoceros-like MAMMALS dating from early Tertiary times. The most spectacular of these had three pairs of horns and a pair of tusks.

DINOFLAGELLATE A marine or freshwater unicellular EUKARYOTE with membrane-bounded nuclei and two flagella of different length, common as planktonic or symbiotic algae; Dinoflagellates arose in the Silurian.

DINOSAUR One of a group of large REPTILES that existed during the Mesozoic era. They are distinguished by the arrangement of hip bones, which were either birdlike (ornithischian) or reptilian (saurischian), and an upright stance.

DIVERGENT EVOLUTION The EVOLUTION of closely related species in different directions, often as a result of diverging lifestyles, eventually leading to the appearance of two very distinct evolutionary lines.

DIVERGENT PLATE MARGIN A LITHOSPHERIC PLATE boundary at which two plates are moving apart and there is upwelling of MANTLE-derived material to create new CRUST. It is associated with MID-OCEAN RIDGES, axial rifts, and active submarine volcanism.

DNA Deoxyribonucleic acid. The main chemical constituent of CHROMOSOMES.

DOLLO'S LAW The evolutionary principle proposed by the Belgian paleontologist Louis Dollo (1857–1931), which states that once a structure is lost or changed it will not reappear in new generations.

DOLOMITE A SEDIMENTARY ROCK consisting of a mineral of the same name, which is a form of magnesium carbonate.

DREIKANTER A rock that has been eroded into a three-sided shape by the wind.

DRUMLIN An elongated hillock composed of sediment deposited by a GLACIER. Its longer axis is parallel to the ice's movement.

DUNE A mound of sand, usually found on a beach or in a desert, built up and driven along by the wind.

DUNE BEDDING The CROSS-BEDDING resulting from the depositional action of wind during dune construction in deserts.

E

ECCENTRICITY (orbital) The degree to which the orbit of an object such as a planet or moon departs from that of a circle.

ECHINODERM One of a group of spiny-shelled DEUTEROSTOME marine INVERTEBRATES, characterized by fivefold radial symmetry, a calcareous internal skeleton or skeletal plates, and hydraulically-powered "tube-feet" equipped with suckers. Echinoderms incorporate the starfish, the brittle stars, the sea urchins, and the crinoids.

ECOLOGY The study of animal and plant communities and their relationships to their surroundings.

ECOSYSTEM An integrated ecological unit consisting of the living organisms and the physical environment (biotic and abiotic factors) in an area. Ecosystems may be largescale or smallscale: the Earth is an ecosystem, and so is a pond. The passage of energy and nutrients through an ecosystem is its FOOD CHAIN.

EDENTATE A member of an order of MAMMALS that encompasses the anteaters, sloths, and armadillos. Edentates lack teeth.

EDIACARA FAUNA An assemblage of late Precambrian fossils known from the Ediacara region of Australia, and consisting of wormlike and seapenlike organisms.

ELEMENT (chemical) A substance that cannot be divided into simpler components and whose atomic structure is constant.

ENDOPTERYGOTE A member of the subclass of insects that have a larval form very different from the adult form. A butterfly with its caterpillar larva and its winged adult is an example. The opposite is an exopterygote.

EOCRINOID An Early Paleozoic stemmed sessile ECHINODERM, a possible ancestor of crinoids.

EPEIRIC SEA An extensive shallow inland sea.

EPICONTINENTAL SEA See EPEIRIC SEA.

EPIPELAGIC An organism inhabiting the upper (PHOTIC) zone of the WATER COLUMN.

ERRATIC BOULDER A boulder that has been transported and deposited by a GLACIER and so is different from the rocks around it.

ESKER A winding ridge of MORAINE left behind by streams flowing beneath a sheet of ice.

EUKARYOTE An organism with a complex cell structure of a clearly defined nucleus containing DNA and separated from the rest of the structure by a membrane, and with specialized organelles such as mitochondria. Eukaryotes include all organisms except bacteria and cyanobacteria, which are PROKARYOTES.

EURYPTERID An aquatic Ordovician–Devonian chelicerate resembling a modern water scorpion.

EVAPORITE A SEDIMENTARY ROCK or bed formed from minerals precipitated from water as a lake, or an inlet of the sea, dries out.

EVOLUTION The process of biological change by which organisms come to differ from their ancestors. The idea of gradual evolution (as opposed to CREATIONISM) gained acceptance in the nineteenth century but remained controversial into the twenty-first because it contradicted many traditional religious beliefs. The British naturalist Charles Darwin (1809–1882) assigned the major role in evolutionary change to NATURAL SELECTION (that is, environmental pressures acting through competition for resources). The current theory of evolution (neo-Darwinism) combines Darwin's theory with the genetic theories of Gregor Mendel and Hugo de Vries' theory of MUTATION. Evolutionary change may have long periods of relative stability interspersed with periods of rapid change (PUNCTUATED EQUILIBRIUM).

EXOPTERYGOTE A member of the subclass of insects that grow by a series of molts, so that the juvenile form is similar to the adult. A newly-hatched grasshopper, for example, is like a miniature version of the adult. The opposite is an ENDTOPTERYGOTE.

EXOSKELETON The hard shell of an insect or similar animal.

EXOTIC TERRANE A comparatively small piece of "foreign" LITHOSPHERE fused to the edge of a continent.

EXTINCTION The complete disappearance of a species or other group of organisms, which occurs when its reproductive rate falls below its mortality rate. Most extinctions in the past occurred because species could not adapt quickly enough to natural changes in the environment; today they are primarily due to human activity.

EXTRUSIVE ROCK An IGNEOUS ROCK, the product of an eruption, that appears on the surface of the Earth as opposed to forming within it.

F

FACIES An assemblage of different SEDIMENTARY ROCKS that are spatially connected and related to the same geological event, so that they represent local conditions.

FAMMATINIAN OROGENY A phase of mountain building in South America during the middle Ordovician, following the accretion of the Andean Precordillera to Gondwana.

FARALLON PLATE A TECTONIC PLATE in the eastern Pacific that was subducted beneath the North American plate during Tertiary times.

FAULT A fracture in a body of rock along which one mass of rock moves against another. Normal faults are caused by extension of the CRUST and may occur in pairs to form GRABEN. Faults also occur along thrusts where there is crustal shortening (reverse faults) and where there is lateral movement of adjacent blocks with little or no vertical slipping (STRIKE-SLIP or TRANSFORM FAULTS).

FELDSPAR Any of a group of alumino-silicate rock-forming minerals.

FELSIC CRUST The light-colored, low-density IGNEOUS ROCKS, with a high FELDSPAR and silica content, that form continents. GRANITE is a felsic rock abundant in CRUST.

FILTER-FEEDER A consumer that strains organic particles or dissolved organic matter from the surrounding water.

FIRN Fallen snow that has partially frozen on top of a GLACIER but has not formed ice.

FLAGELLATE A collective name for unicellular organisms that move by means of flagella.

FLOWERING PLANT *See* ANGIOSPERM.

FLYSCH Thick deposits of SANDSTONE and SHALE eroded from newly-arisen mountain ranges. The term is sometimes restricted to deposits to the north and south of the Alps.

FOLD Originally horizontal strata of SEDIMENTARY ROCK that have been warped and bent by mountain-building. Folds may form mountain massifs by compressional deformation, as in the case of the Alps and the Appalachians.

FOLD BELT A long, narrow zone of CRUST in which there has been intense deformation and development of FOLDS. Such belts usually develop along continental margins associated with CONVERGENT PLATE MARGINS. They have also been recognized on Venus.

FOLD-AND-THRUST BELT The inland zone of a mountain chain, characterized by FOLDS and THRUST FAULTS.

FOOD CHAIN The levels of nutrition in an ecosystem, beginning at the bottom with primary PRODUCERS, which are principally plants, to a series of CONSUMERS—herbivores, carnivores, and decomposers.

FOOD WEB A more complex food chain, with several species at each level, so that there is more than one PRODUCER and more than one CONSUMER of each type.

FORAMINIFERAN An order of single-celled PROTOZOANS, mostly marine, with tests (shells) usually of CALCIUM CARBONATE and perforated by pores and reinforced with minerals. They evolved in the Cambrian.

FOREARC BASIN An elongate BASIN of deposition that lies between an ISLAND ARC and the wedge of ACCRETION behind it.

FOREDEEP The deep area on the oceanic side of an ISLAND ARC.

FORMATION The basic stratigraphic unit: a body of rock with distinctive lithographic features that can be mapped.

FOSSIL The remains of a once-living creature found preserved in rocks. Fossils may be part of the creature, or the shape of the creature turned to stone, or even just traces such as footprints or worm burrows.

FRAGMENTATION A process of splitting of a continent into smaller pieces during RIFTING and following spreading.

FUNGUS A EUKARYOTE that forms spores as slender septate tubes (hyphae) and lacks any motile stage during its life cycle. Fungi have existed since the Proterozoic.

FUSULINIDS A group of calcareous coiled FORAMINIFERANS, abundant during the Carboniferous and Permian periods, many of which were spindle-shaped.

G

GABBRO A coarse-grained IGNEOUS ROCK similar to BASALT in composition but formed below the Earth's surface. It contains FELDSPAR.

GAMETE The reproductive cell of an organism that fuses with one from another organism during sexual preproduction.

GASTROPOD A snail: a univalved MOLLUSK possessing a well-developed head and foot, and asymmetrical development of its shell and inner organs. Gastropods appeared in the Cambrian.

GENE The basic unit of inheritance that controls a characteristic of an organism. It may be considered as a length of DNA organized in a very specific manner. Genes mutate and recombine to produce the variation on which NATURAL SELECTION acts.

GENE POOL The mix of genetic material in breeding populations of organisms.

GENUS The sixth division under the Linnaean system of classification for living things, consisting of a number of similar or closely related species. Similar genera are grouped into families.

GEOSPHERE The solid portion of the Earth, as distinct from the ATMOSPHERE or BIOSPHERE.

GEYSER An eruption of hot water from the ground. Groundwater boils below the surface by volcanic heat, and the expanding superheated steam pushes water out of a vent. This releases the pressure and the rest of the water column boils explosively, blasting the water high into the air.

GLACIAL MAXIMUM A period during an ICE AGE in which the glaciation is most extensive.

GLACIER A thick mass of ice and compacted snow that moves slowly downhill and persists year-round for up to 100 years.

GLACIOEUSTASY Fluctuations in global sea levels caused by changes in the quantity of sea water as ice caps grow or melt.

GLOBAL WARMING A change in the Earth's climate involving an increase of overall temperatures, sometimes as a result of natural processes, but mostly attributed in the present to the GREENHOUSE EFFECT. Such a change is not regular and may show itself as unpredictable weather conditions caused by the increased temperature gradient between permanent ice caps and the warmer conditions of the rest of the world. The United Nations Environment Program estimates that by the year 2025 global warming will cause average world temperatures to rise by 2.7°F (1.5°C), with a consequent rise of 8in (20cm) in sea level caused by the melting of polar ice.

GLOSSOPTERIS A type of seed-bearing fern that characterized the BIOGEOGRAPHIC PROVINCE of GONDWANA at the end of the Paleozoic.

GNEISS A coarse-grained METAMORPHIC ROCK that exhibits alternating bands of dark and light materials. Such rocks have formed at depth within MOBILE BELTS inside the Earth. Granitic gneisses are frequently associated with granites in continental CRATONS.

GONDWANA The supercontinent incorporating today's southern continents—Africa, South America, Australia, and Antarctica—and also India, Madagascar, and New Zealand during the Paleozic and Mesozoic.

GONIATITE A member of a group of CEPHALOPODS similar to the modern NAUTILUS, with a characteristic jagged pattern of suture lines.

GRABEN A geological structure in which a section of rock has descended between parallel FAULTS. On the surface this may show up as a RIFT VALLEY.

GRANITE A hard, coarse-grained IGNEOUS ROCK that consists mainly of QUARTZ and FELDSPAR, often with mica or other colored minerals. Most regions of granite have resulted from the crystallization of molten MAGMA, though some granites may have formed through the metamorphism of other, existing rocks. The extrusive equivalent is rhyolite.

GRAPTOLITE A Cambrian–Carboniferous filter-feeding colonial HEMICHORDATE. Graptolites lived near the surface of the oceans, like modern PLANKTON, and are known from fossils of their simple outer skeletons. Most died out at the end of the Silurian period.

GRASSES A group of Cenozoic ANGIOSPERM plants characterized by long straplike leaves and underground stems.

GRAZER In a TROPHIC WEB, a herbivorous CONSUMER that crops along the substrate.

GREENHOUSE EFFECT The gradual increase of the temperature of air in the lower ATMOSPHERE, which is believed to be due to the buildup of gases such as carbon dioxide, OZONE, methane, nitrous oxide, and CHLOROFLUOROCARBONS. These trap solar radiation absorbed and re-emitted from the Earth's surface, preventing it from escaping into space and thus overheating the Earth.

GREENSTONE A SEDIMENTARY ROCK formed on the Earth's surface in Precambrian times when the atmosphere was deficient in oxygen. The green color is derived from minerals that form under low-oxygen conditions.

GREYWACKE A poorly sorted, dark, very hard, coarse-grained SEDIMENTARY ROCK with angular particles.

GULF STREAM The ocean current that travels from the Gulf of Mexico northeast across the North Atlantic—where it becomes the North Atlantic Drift—bringing warm conditions to the west coast of Europe.

GYMNOSPERM An informal designation for a seed-bearing plant that does not protect its seeds in an enclosed fruit. Conifers and ginkgoes are examples.

H

HADROSAUR A member of a group of ORNITHOPOD dinosaurs of the Cretaceous, characterized by a broad ducklike bill.

HAECKEL'S LAW An evolutionary principle, now modified, that states that the young of a species resemble the adults of their ancestors (ontogeny recapitulates PHYLOGENY).

HALKIERIID An Early Paleozoic COELOSCLERITOPHORAN with two shells, a sluglike foot, and an upper surface covered with spiny and scalelike SCLERITES.

HANGING VALLEY A side valley that enters a glaciated U-SHAPED VALLEY some way up the valley side.

HEMICHORDATE A primitive wormlike marine DEUTEROSTOME with a NOTOCHORD. The body is subdivided into a head shield, collar, and trunk perforated by gill slits. Hemichordates appeared in the Cambrian.

HERBIVORE Any animal that eats plants. Unlike CARNIVORE, the term is not restricted to any particular group of animals.

HERCYNIAN OROGENY A mountain-building event during the Late Paleozoic that emplaced many of the granitic masses now found in western Europe. North America's equivalent was the ALLEGHENIAN OROGENY.

HETEROTROPH An organism that is a CONSUMER —one that takes in ORGANIC compounds as food because it cannot synthesize them itself from simple inorganic substances. Heterotrophs include bacteria, fungi, protozoa, and all animals. Plants, which are AUTOTROPHS (producers), are the main exception.

HEXACORAL One of a group of tentacle-feeding CORALS, characterized by their hexagonal symmetry, that replaced the Paleozoic RUGOSAN corals during the Mesozoic. Hexacorals still live today.

HIRNANTIAN ICE AGE The first Phanerozoic glaciation, which occurred at the very end of the Ordovician (Hirnantian stage).

HOMINID A late HOMINOID, characterized by its upright gait (bipedalism), and a member of the line from which modern humans evolved about 4 or 5 million years ago.

HOMINOID A PRIMATE such as one of the lesser apes (gibbons and siamangs), great apes (orangutans, gorillas, and chimpanzees), and humans.

HOMO The genus to which humans belong. Two species are recognized besides the modern *H. sapiens*: *H. habilis*, the first tool-maker, and *H. erectus*, which was the first to spread around the world from Africa.

HOMOLOGOUS STRUCTURE A structure that is similar in different species, suggesting a common ancestor, but serves a different function in each, such as arms and wings.

HORST BLOCK A geological structure, the reverse of a GRABEN, in which a section of rock is raised between two FAULTS. The surface feature of this may be a flat-topped hill.

HOT SPOT A place at which a MANTLE PLUME causes hot MAGMA to rise toward the base of the Earth's CRUST, producing high heat flow and volcanism at the surface. Iceland and the Hawaiian islands lie above hot spots.

HYDROSPHERE The part of the Earth's structure made up of water. This includes the oceans, the ice caps, and the gases in the ATMOSPHERE.

HYDROTHERM A deep, hot (900° to 7200°F/ 500° to 4000°C) marine spring of mineral waters, often the site of a BLACK SMOKER.

HYOLITH A marine bivalved Paleozoic INVERTEBRATE, a possible relative of BRACHIOPODS.

HYPSILOPHODONT One of a group of small ORNITHOPOD dinosaurs built for fast running.

I

IAPETUS OCEAN The ocean that existed between LAURENTIA, AVALONIA, and BALTICA before they came together to form Euramerica; it is sometimes known as the Proto-Atlantic Ocean because it lay between what is now North America and Europe.

ICE AGE An extended period of cold climate during which there is an increase in the surface area of GLACIERS and ice caps. There have been several ice ages in the Earth's history, with the most recent occurring in the Pleistocene epoch from 1.8 million years ago to about 12,000 years ago.

ICE SHEET A continental GLACIER that spreads out from an area of extreme cold rather than downhill from a mountain. Antarctica and Greenland are covered by ice sheets.

ICHNOLOGY The study of FOSSIL footprints and burrows.

ICHTHYOSAUR One of a group of marine REPTILES of the Mesozoic that resembled dolphins in appearance.

IGNEOUS CORE The solidified molten material in the middle of a mountain range, formed at very high temperatures.

IGNEOUS ROCK Any rock formed by the solidification of molten MAGMA. There are two main types: intrusive igneous rocks that formed underground, and extrusive igneous rocks that formed from lava that erupted at the Earth's surface. The former, such as GRANITE, are coarse-grained, while the latter, such as BASALT, are fine-grained.

IGUANODONTID One of a group of plant-eating ORNITHOPOD dinosaurs.

INDEX FOSSIL A FOSSIL whose presence indicates a rock's age. Useful index fossil species are short-lived and widespread; GRAPTOLITES and AMMONITES are examples. Index fossils are also known as zone fossils.

INDUSTRIAL REVOLUTION The rise of the use of machinery in industry, which began in the late eighteenth century in Britain.

INSECT One of a group of air-breathing ARTHROPODS that appeared in the Devonian. The body is subdivided into a head, a chest with three pairs of legs, an abdomen, and one or two pairs of wings.

INTERGLACIAL A period of milder climates occurring within an ICE AGE.

INVERTEBRATE An animal with no backbone—95 percent of all animal species.

ISLAND ARC A chain of volcanic islands that develops at the edge of an ocean trench, where volcanoes are generated by the melting of a subducting plate (see SUBDUCTION).

ISOSTASY The balance between parts of the Earth's surface caused by differences in density, based on the principle that the rocks of the CRUST "float" on those of the underlying MANTLE. Ocean crust is made from dense BASALT, whereas upper continental crust is mostly low-density FELSIC rocks with deep "roots" to compensate for its lightness.

ISOTOPE One of a number of forms of a chemical ELEMENT that has the same number of protons in its nucleus but a different number of neutrons and thus different physical properties.

J

JASPER A METAMORPHIC ROCK formed chiefly on siliceous deep-marine sediments.

JAWLESS FISH A fishlike CRANIATE lacking jaws, trunk bones, and, in most cases, paired fins. Jawless fish appeared in the Ordovician.

JUAN DE FUCA RIDGE An ocean ridge off the west coast of Canada, an isolated part of the East Pacific Rise that is now being subducted beneath the North American Plate.

K

KARST A LIMESTONE landscape, characterized by extreme dryness and EROSION of the exposed rocks into blocks (CLINTS) separated by deep gullies (GRYKES), caused by chemical breakdown of CALCITE in the limestone.

KERGUELEAN LANDMASS A submerged continental plateau in the southern Indian Ocean.

KETTLE HOLE A depression formed in an area of MORAINE—rocky debris left by a retreating GLACIER. A block of glacier ice left stranded eventually melts, leaving this structure.

KOMATIITE An extrusive IGNEOUS rock composed of peridotite that was widespread in Archean times and was the precursor of BASALT rocks in the Earth's crust.

KT BOUNDARY EVENT The boundary between the Cretaceous and Tertiary periods about 65 million years ago, marked by a MASS EXTINCTION in which the DINOSAURS perished, among others. One possible explanation is a METEORITE strike on the Earth, which left a 110mi (180km) crater in the Gulf of Mexico.

L

LAGERSTATTEN Localities containing FOSSILS preserved much better than is usual.

LAMARCKISM A theory of EVOLUTION proposed by the French biologist Jean-Baptiste de Lamarck (1744–1829), by which traits acquired by an individual during its lifetime could be passed down to offspring. It was discredited by the work of Charles Darwin.

LARAMIDE OROGENY A mountain-building event that took place in western North

America in late Cretaceous times and contributed to the Rocky Mountains.

LAURASIA The supercontinent that represented the northern part of PANGEA, consisting of the landmasses that now constitute North America, Europe, and northern Asia.

LAURENTIA The supercontinent that broke up to form North America and parts of Europe.

LAVA Molten rock material that rises from the Earth's interior, as in a volcanic eruption. BASALT is a typical lava.

LEPTOSPORANGIATE FERNS "True" ferns. The general term "ferns" includes several groups of similar appearance which are not MONOPHYLETIC in origin.

LICHEN A symbiotic organism consisting of a FUNGUS and CYANOBACTERIUM or green alga; only modern species are known.

LIGNITE A soft brown form of COAL.

LIMESTONE A CARBONATE SEDIMENTARY rock consisting primarily of CALCITE, which may be derived from solution in seawater or by the accumulation of animal shells.

LITHIFICATION The hardening of sediment until it forms rock.

LITHISTID DEMOSPONGE A marine DEMOSPONGE with a rigid skeleton. Lithistids have existed since the Cambrian.

LITHOSPHERE The solid outer layer of the Earth, about 60mi (100km) deep, consisting of the CRUST and the uppermost part of the MANTLE. It is segmented into plates floating on the more fluid ASTHENOSPHERE.

LITOPTERN One of a group of extinct South American UNGULATES, some of them horselike, from Tertiary times.

LOBE-FINNED FISH A bony fish belonging to the subclass sauropterygii. Lobefins are distinguished from ray-finned fish by the fleshy lobes that support the fins. They are considered ancestors of the AMPHIBIANS and hence of all land VERTEBRATES.

LOWLAND A part of land where accumulation processes prevail over destruction.

"LUCY" The first known FOSSIL skeleton of the early human ancestor *Australopithecus afarensis*: an adult female AUSTRALOPITHECINE discovered in Ethiopa in 1974.

LYCOPOD/LYCOPSID See CLUBMOSS.

M

MAFIC CRUST The relatively heavy rock material that underlies the oceans. The principal constituents are magnesium and FELDSPAR.

MAGMA Molten rocky material that forms in the Earth's lower crust or MANTLE. When it solidifies, it is known as IGNEOUS ROCK; when it erupts at the surface, it is LAVA.

MAMMAL Any member of the vertebrate class Mammalia, which contains about 4000 species. Their most distinctive characteristic is mammary (milk) glands in the female. There are three orders: PLACENTAL, MARSUPIAL,

and MONOTREMES. Placentals are the most common, monotremes the least.

MANICOUAGAN EVENT A meteorite impact in Quebec, Canada, at the end of the Triassic.

MANTLE The section of the Earth's structure that lies between the thin outer crust and the core. At almost 1800mi (2900km) thick, the mantle comprises the greatest part of the Earth's volume. Like that of other terrestrial planets, it is composed of dense silicates of iron and magnesium, whereas the mantles of the gaseous outer planets are thought to be mostly hydrogen.

MANTLE PLUME An upwelling lobe or jet of hot, partly molten material rising within the Earth's mantle. Plumes are believed to give rise to volcanic islands away from the edges of continental plates, such as Hawaii.

MARGINAL SEA A semi-closed sea attached to a continent and formed during RIFTING and early spreading.

MARSUPIAL A member of an order of MAMMALS that nurture their immature young in a pouch. Kangaroos and wombats are among the few modern marsupials, but in Tertiary times the group was widespread.

MASS EXTINCTION The EXTINCTION of a significant proportion of the Earth's organisms over a wide area in a short span of time.

MEGALONYCHID A member of a group of extinct giant ground sloths dating from the late Tertiary and Quaternary.

MESONYCHID A member of a group of primitive omnivorous UNGULATE MAMMALS dating from the early Tertiary. They included the wolf-sized *Mesonyx* and the giant *Andrewsarchus*.

MESOPELAGIC The middle zone of the WATER COLUMN and the organisms that inhabit it.

MESOZOIC The era of geological time from 248 to 65 million years ago, encompassing the Triassic, Jurassic, and Cretaceous periods.

MESSINIAN CRISIS The biological disruption caused by a lowering of sea levels at the end of the Miocene. The Antarctic ice sheets expanded and the Mediterranean dried up.

METAMORPHIC BELT An elongated region of METAMORPHIC ROCKS exposed at the core of an ancient FOLD mountain chain.

METAMORPHIC ROCK A rock, usually SEDIMENTARY in origin, that is subjected to such heat or pressure that it recrystallizes into new minerals without leaving the solid state. If it melted at any time, the result is an IGNEOUS ROCK.

METAZOAN Any multicellular animal whose cells are organized in tissues—that is, all animals except PROTOZOANS. Metazoans appeared in the latest Proterozoic.

METEORITE A piece of interstellar rocky debris, particularly one that has fallen to Earth.

MIACID A member of a group of primitive carnivorous MAMMALS from the early Tertiary, which diversified into VULPAVINES and VIVERRAVINES—the dog and cat branches.

MICROPLATE A small LITHOSPHERIC plate, usually composed mainly of FELSIC rocks.

MID-OCEAN RIDGE The raised feature on the ocean floor where new OCEANIC CRUST emerges, spreading laterally on each side. It forms a long ridge with volcanoes, HYDROTHERMS, and RIFT VALLEYS along the crest. Such ridges are frequently offset by TRANSFORM FAULTS which assist the MANTLE's attempts to "bend" the brittle margins of LITHOSPHERIC PLATES. Ridges may develop above CONVECTION cells in the mantle.

MILANKOVITCH CYCLE The cycle of changes in the Earth's movements (ECCENTRICITY of the orbit, PRECESSION of the rotational axis, and obliquity), invoked first as an explanation for ICE AGES in the late eighteenth century, and revived by Milutin Milankovitch.

MINERAL A naturally formed inorganic chemical substance with a particular chemical composition. A rock consists of crystals of several different types of mineral.

MINERALIZED SKELETON A skeleton made of minerals: mostly CARBONATES, phosphates, and silicon oxides.

MITOCHONDRIA A microscopic structure in the living cells of PROKARYOTES that provides energy for the cell's working. It may be a descendant of small bacteria that were trapped within larger ones.

MITOCHONDRIAL DNA (mtDNA) DNA found in mitochondria. It evolves more rapidly than nuclear (ordinary) DNA, so it can be used to trace the divergence of populations, and is passed down only through the maternal line (see MITOCHONDRIAL EVE). The small variation in mtDNA found in humans supports the out of Africa hypothesis of human origins.

MITOCHONDRIAL EVE The nickname for the hypothetical female ancestor who was the source of all MITOCHONDRIAL DNA in humans today.

MOBILE BELT A region of intense geological activity located along a plate margin. Mobile belts are characterized by volcanism, seismic activity, and mountain-building.

MOHOROVICIC DISCONTINUITY (Moho) The boundary between the Earth's crust and its mantle at which the speed of seismic waves increases sharply. The depth of the Moho varies, from about 6mi (10km) below the ocean floor to 20mi (35km) below continents and 40mi (70km) below mountains.

MOLASSE An assemblage of usually coarse-grained, non-marine SEDIMENTARY ROCKS formed by rapidly eroding new mountains.

MOLLUSK Any of the INVERTEBRATE phylum Mollusca, which appeared in the Cambrian.

MONOPLACOPHORAN An early single-valved marine MOLLUSK possessing a caplike, bilaterally symmetrical calcareous shell.

MONOPHYLETIC A group that contains all the descendants of a single common ancestor.

MONOTREME A member of an order of MAMMALS that lay eggs. The echidna and platypus are the only living monotremes.

MORAINE Rocky debris picked up by a GLACIER, carried along and deposited elsewhere.

MUDSTONE A fine-grained SEDIMENTARY ROCK formed by the consolidation of mud. It resembles SHALE but lacks distinct fine bedding.

MULTITUBERCULATE A member of an order of primitive rodentlike MAMMALS from the Mesozoic and early Tertiary. They may have been the first herbivorous mammals.

MUTATION A change in the genetic makeup of an organism produced by an alternation in its DNA. Mutations, the raw material of EVOLUTION, result from mistakes during replication (copying) of DNA. Only beneficial mistakes are therefore favored by NATURAL SELECTION.

MYSTICETE A BALEEN whale.

N

NANOPLANKTON Microscopic plankton.

NATURAL SELECTION The primary mechanism of EVOLUTION, first stated by Charles Darwin, by which gene frequencies in a population change through certain individuals producing more descendants than others. Because most environments are slowly but continuously changing, natural selection enhances the reproductive success of individuals that possess favorable characteristics. The process is slow, relying on random variation in the genes of an organism due to MUTATION and on genetic recombination during sexual reproduction.

NAUTILOID A member of a subclass of CEPHALOPODS with coiled shells, related to the AMMONITES and GONIATITES. Abundant in the Early Paleozoic, they are all but extinct.

NEANDERTAL A member of the subspecies *Homo sapiens neanderthalensis*, closely related to modern humans (*H. sapiens sapiens*), preceding them for much of the Pleistocene. Named after Germany's Neander valley where they were discovered.

NEKTONIC An organism that actively swims in the WATER COLUMN, as opposed to floating.

NEOLITHIC The "New Stone Age," a culture characterized by the use of advanced stone tools and the development of agriculture around the end of the Pleistocene ice age.

NEVADAN OROGENY The mountain-building episode along the west coast of North America during the Jurassic and Early Cretaceous periods, which contributed to the Western Cordillera.

NEW RED SANDSTONE The sequence of terrigenous sedimentary rocks deposited in the supercontinent of Laurasia during the Permian and Triassic periods.

NEW WORLD MONKEYS *See* PLATYRRHINES.

NICHE The ecological role of a species; the total combination of environmental factors to which a species is adapted. In biological terms, the space in a particular environment that is occupied by a particular creature on account of its lifestyle.

NONCONFORMITY A type of stratigraphic UNCONFORMITY that separates a bed of rocks from crystalline rocks below.

NON-SEQUENCE A gap in the stratigraphic succession which arises because deposits from a particular time were never laid down, or because these deposits have subsequently eroded completely. The existence of a gap is proven by other paleontological evidence.

NOTHOSAUR A member of a group of swimming reptiles from the Triassic, and a forerunner of the PLESIOSAURS.

NOTOCHORD A flexible support along the length of the body of certain wormlike animals. The notochord is ancestral to the vertebral column in VERTEBRATES.

NOTOUNGULATE A primitive South American hoofed MAMMAL. *See also* UNGULATE.

NUEE ARDENTE A fast-moving "glowing avalanche" consisting of hot volcanic ash, fine dust, molten lava fragments, and hot gases associated with a volcanic eruption.

O

OBIK SEA A shelf sea that existed in the early Tertiary across part of Russia to the east of the Ural mountains.

OCEAN RIDGE *See* MID-OCEAN RIDGE.

OCEAN TRENCH The deepest part of an ocean, an elongated depression pulled down as one plate slides beneath another during the process of PLATE TECTONICS. ISLAND ARCS usually form at the rim of an ocean trench.

OCEANIC CRUST The relatively heavy basaltic rock, an average of 5mi (8km) thick, that underlies the oceans. Its main constituents are magnesium and feldspar, and the bottom layer gives way to GABBRO and perioditic rocks at the MOHOROVIC DISCONTINUITY.

ODONTOCETE A toothed whale.

OIL SHALE A fine-grained SEDIMENTARY ROCK, formed from the lithification of mud, rich in organic matter, easily split into fine layers or flakes, and combustible.

OLD RED SANDSTONE The sequence of terrigenous SEDIMENTARY ROCKS deposited during the Devonian period from the newly arisen Acadian–Caledonian mountain range, forming thick repeating beds of SANDSTONE.

OLDOWAN CULTURE A stone-tool-making culture that existed in Africa's OLDUVAI GORGE (Tanzania) in the early Pleistocene.

OLDUVAI GORGE A site in Tanzania, in the East African Rift Valley, where there have been many important finds of fossil HOMINIDS since the 1970s, including "LUCY."

OMOMYID A member of a family of primitive lemurs from early Tertiary times.

ONYCHOPHORAN Common name velvet worm: a terrestrial segmented invertebrate with numerous rigid telescopic legs. Velvet worms appeared in the Carboniferous.

OOLITE LIMESTONE formed by tiny particles of CALCITE precipitated out of seawater.

OPHIOLITE An assemblage of rocks—mostly BASALT, GABBRO, and JASPER—representing the remains of OCEANIC CRUST pushed onto land during continental collision and OROGENY.

ORDER A grade of zoological classification. A class may contain several orders, and an order several families. The primates are an order within the class Mammalia and contain several families such as the Omomyidae and the Hominidae.

ORDOVICIAN RADIATION A rapid increase of animal diversity, biomass, and size, as well as the appearance of new groups of CORALS, BRYOZOANS, BRACHIOPODS, TRILOBITES, OSTRACODES, and other INVERTEBRATES and CHORDATES, in the Early and Middle Ordovician.

ORGANIC Anything that contains carbon, except CARBONATES and the oxides of carbon; thus, organic substances include all living things and their products.

ORIGINAL HORIZONTALITY, PRINCIPLE OF The principle of geology that states that all strata are laid down horizontally.

ORIGINAL LATERAL CONTINUITY, PRINCIPLE OF The principle of geology that states that similar rock strata separated by any erosional feature, such as a valley, were originally laid down together.

ORNITHISCHIAN The herbivorous "bird-hips," one of two major groups of DINOSAURS—though not the line from which ARCHAEOPTERYX and the birds evolved.

ORNITHOPODA A line of Jurassic–Cretaceous two-footed plant-eating dinosaurs descended from the ORNITHISCHIANS. It included *Camptosaurus, Hadrosaurus,* and *Iguanodon*.

OROGENY An episode of mountain-building.

OSTEICHTHYAN A jawed fish whose skeletal cartilage is ossified partly or completely. Osteichthyans appeared in the Devonian.

OSTEOSTRACAN A member of a group of early jawless fish, with well-defined skeletons and paired fins, dating from the early Paleozoic.

OSTRACODE A microscopic aquatic CRUSTACEAN that appeared in the Ordovician.

OUT OF AFRICA HYPOTHESIS The widely accepted theory that human beings evolved in Africa and then spread throughout the world, as opposed to their evolving from an already widespread ancestral stock (the little- accepted "multiregional hypothesis").

OVULE A reproductive structure in seed plants that develops into a seed once fertilized.

OZONE LAYER A layer of the atmosphere, particularly rich in ozone gas, that filters out ultraviolet rays from the sun, preventing GLOBAL WARMING and the GREENHOUSE EFFECT. It is vulnerable to atmospheric pollution.

P

PACK ICE Blocks of floating ice compacted together to form a solid surface on the sea.

PALEOASIAN OCEAN The ocean that separated Siberia and eastern GONDWANA in the latest Proterozoic and early Paleozoic.

PALEOLITHIC The "Early Stone Age," an Early Pleistocene culture represented by the earliest and most primitive stone tools. *See* OLDOWAN CULTURE.

PALEOMAGNETISM The study of the condition of the Earth's magnetic field and its properties in the geologic past. This magnetic field leaves an impression in the rock formed at the time, and this gives clues as to the position of poles and continents in history.

PALEOTHETIS OCEAN A body of water, forerunner of the THETIS OCEAN, that existed as a vast embayment into PANGEA almost separating LAURASIA from GONDWANA in the middle and late Paleozoic.

PALEOZOIC The era of geological time from 545 to 248 million years ago, encompassing the Cambrian, Ordovician, and Silurian periods (the early Paleozoic) and the Devonian, Carboniferous, and Permian periods (the late Paleozoic).

PALEOZOIC FAUNA The animals produced by the ORDOVICIAN RADIATION and subsequent diversification. Most of them (TRILOBITES, GRAPTOLITES, and others) vanished by the end of the Paleozoic, but a few (CEPHALOPODS, ECHINODERMS) persisted into modern times.

PANGEA The late Paleozoic–early Mesozoic supercontinent comprised of every major continental landmass.

PANTHALASSA OCEAN The single ocean that covered the northern hemisphere in the Paleozoic and early Mesozoic. It was a predecessor of the Pacific Ocean.

PARAPHYLETIC A group evolved from more than one ancestor (the opposite of MONO-PHYLETIC). The dinosaurs may be paraphyletic because the two major groups (saurischians and ornithischians) may have evolved independently.

PARATETHYS The shallow sea in the area of the Black and Caspian seas in late Tertiary times.

PAREIASAUR A member of a group of herbivorous REPTILES from the Permian period. They were big heavy animals and may have been closely related to the ancestors of the turtles.

PECORAN A member of the group of even-toed UNGULATES encompassing deer and giraffes.

PELAGIC Describing an organism that inhabits the open sea, including forms that are free-swimming (NEKTONIC) and those that float passively (PLANKTONIC).

PELLET CONVEYOR A natural water-purification system that evolved in the Cambrian, when microscopic ZOOPLANKTON began to remove the organic waste of other animals from the top of the water to the seabed, where DETRITUS FEEDERS could make use of it.

PELYCOSAUR A member of the most primitive group of mammal-like REPTILES, many of which had sails on their backs.

PENTASTOME A parasitic CRUSTACEAN that appeared in the Cambrian. It is also known as the tongue worm for its flat soft body covered by a soft ringed cuticle.

PERISSODACTYL An odd-toed UNGULATE.

PERMAFROST The permanently frozen soil and subsoil in the Arctic and sub-Arctic regions of the Earth.

PETROLEUM Crude oil, which forms from decayed ORGANIC matter found concentrated in rock structures called traps, and is extracted as the raw material for industry.

PHANEROZOIC The geological eon from the beginning of the Paleozoic onwards, when the first identifiable fossils formed.

PHOSPHORITE A SEDIMENTARY ROCK consisting of phosphate minerals as guano, shells, or bacterial deposit.

PHOTIC ZONE The depth to which light penetrates a body of water sufficiently to permit PHOTOSYNTHESIS—up to about 300ft (100m).

PHOTOSYNTHESIS The process by which a plant extracts the energy of sunlight and uses it to manufacture food from water and carbon dioxide in the atmosphere.

PHYLOGENY The sequence of changes that occurs in a given species or other taxonomic group during the course of EVOLUTION.

PHYLUM A category of organisms that consists of one or more similar or closely related classes. Related phyla are grouped together in a kingdom. Chordata and Mollusca are two examples of phyla.

PHYTOPLANKTON An algal PLANKTON. Phytoplankton consist mostly of ALGAE and carry out almost all photosynthesis in the oceans. They are the basis of the FOOD CHAIN.

PINNIPED A member of the group of carnivorous MAMMALS that encompasses the seals and walruses.

PLACENTAL A member of the order of MAMMALS that nurture their young in the uterus before birth. This encompasses all modern mammals except the MARSUPIALS and the egg-laying MONOTREMES.

PLACODERM A Silurian–Devonian jawed cartilaginous fish with a head covered by rigid bony shields.

PLACODONT One of a group of marine REPTILES, mostly slow-moving shellfish-eaters and some with turtle-like shells, from Triassic times.

PLANKTON Small, often microscopic free-floating organisms that inhabit the top layer of the water column and are an important food source for larger animals. They include PHYTOPLANKTON and ZOOPLANKTON.

PLATE TECTONICS The theory that invokes the movement and interaction of LITHOSPHERIC PLATES as an explanation for CONTINENTAL DRIFT, SEAFLOOR SPREADING, volcanism, earthquakes, and mountain-building.

PLATYRRHINE A member of the NEW WORLD MONKEY group, characterized by a narrow nose and usually prehensile tail, which were not involved in the evolution of humans (*see also* CATARRHINES).

PLAYA An enclosed flat BASIN in a desert, usually occupied in part by an ephemeral (seasonal) lake or lakes. When these dry out, they form EVAPORITE deposits.

PLESIOSAUR A carnivorous swimming REPTILE of the Mesozoic. Plesiosaurs had turtle-like bodies and paddles and long necks.

PLIOSAUR One of a group of large-headed, short-necked Mesozoic swimming REPTILES.

PLUME *See* MANTLE PLUME.

PLUTON A body of intrusive IGNEOUS ROCK that has formed beneath the Earth's surface.

PLUVIAL LAKE A lake formed by rain.

POLAR WANDERING The slight changes in position of the Earth's magnetic poles due to CONTINENTAL DRIFT and PLATE TECTONICS.

POLARITY REVERSAL The reversal of the Earth's magnetic field at intervals of 10,000 to 25,000 years, leaving magnetic stripes in seafloor rocks. *See* PALEOMAGNETISM.

POLYCHAETE A mostly marine ANNELID worm that appeared in the Cambrian. Each segment has a pair of fleshy flaps bearing a bundle of bristles (setae).

POLYPLACOPHORAN A multivalved marine MOLLUSK possessing a bilaterally symmetrical, calcareous multiple shell. Polyplacophorans evolved during the Cambrian.

PRECESSION In astronomy, the apparent slow motion of the celestial poles, largely due to the wobbling of the Earth's rotational axis induced by the gravitational pull of the Sun and Moon. The axis gradually changes direction over a cycle of about 26,000 years; this is why the equinoxes occur earlier each succeeding year. *See* MILANKOVITCH CYCLE.

PRECORDILLERA A South American terrane that split from the Appalachian margin of Laurentia (future North America) during the Cambrian. The Andes later formed there.

PREDATOR An animal that kills and eats others.

PRIMATE A member of the highly derived order of MAMMALS that contains the lemurs, the monkeys, the apes, and humans.

PROBLEMATIC FOSSIL A fossil of an organism that has no apparent affinity with any modern phylum. Problematic fossils are especially abundant in Cambrian strata.

PRODUCER An organism that produces organic matter by modifying light (PHOTOSYNTHESIS) or chemicals (chemosynthesis).

PROKARYOTE A very simple cell in which the genetic material is not confined to a nucleus but spread through the cell structure. Only primitive bacteria and cyanobacteria are prokaryotes; all other organisms are EUKARYOTES.

PROSIMIAN One of the basal members of the PRIMATE order, including tarsiers, lemurs, and tree shrews.

PROTEIN A complex organic compound made up of AMINO ACIDS and forming the bulk of a living thing.

PROTOCTIST A member of the kingdom Protoctista (sometimes called Protista), a category for all organisms that are neither bacteria, animals, plants (nor fungi, according to some scientists): that is, they are algae, protozoa, and slime molds (and some fungi).

PROTOSTOME "First mouth": an animal in which the embryonic mouth develops into the adult mouth. It includes most bilaterally symmetrical invertebrates.

PROTOZOAN A unicellular organism, one of the earliest EUKARYOTES.

PSILOPHYTE An old name for ancient VASCULAR PLANTS. It now refers to a number of primitive TRACHEOPHYTES of different origins, namely rhyniophytes, zosterophyllophytes, and trimerophytes.

PSYCHROSPHERE The near-freezing waters of the deepest part of the ocean, formed by CONVECTION, which causes the cold water at the poles to sink.

PTEROBRANCH A mostly colonial dendritic sessile HEMICHORDATE from the Cambrian.

PUNCTUATED EQUILIBRIUM A pattern of EVOLUTION in which periods of comparative stability are interspersed with bursts of increased VARIATION and the formation of new species. The duration of the respective periods varies greatly under different environmental circumstances.

PYCNOGONID (sea spider) An articulated marine INVERTEBRATE, with a thin body and jointed legs, that appeared in the Devonian.

PYGIDIUM The tail portion of a TRILOBITE.

PYRITE A gold-yellow mineral of iron sulfide, and an important source of sulfur and iron.

PYROCLASTIC A SEDIMENTARY ROCK consisting of fragments of volcanic material.

Q

QUARTZ One of the most widespread SILICATE minerals in the Earth's CONTINENTAL CRUST and the chief component of CLASTIC SEDIMENTARY rocks. It is mostly silicon dioxide (SiO_2).

QUATERNARY The era of geological time that encompasses the Pleistocene and the Holocene epochs, and so covers the last ICE AGE and the whole of human history.

R

RADIATION (ADAPTIVE) The process by which a lineage evolves different forms, allowing members to adapt to different lifestyles in different environments.

RADIOACTIVE DECAY The process in which a radioactive element sheds neutrons and changes its atomic number, thereby becoming a completely different substance.

RADIOCARBON DATING A type of RADIOMETRIC DATING using carbon–14, which has a very short half-life, and can be used to date younger rocks (up to about 70,000 years).

RADIOCYATH An Early Cambrian RECEPTACULITID with multi-rayed heads.

RADIOMETRIC DATING The technique used to estimate the age of a rock or mineral by calculating how much of its radioactive matter has decayed since it was formed.

RAISED BEACH A coastal landform consisting of a flat shelf at some height above sea level, indicating the position of sea level at some time in the past. Raised beaches are a product of glaciation and the weight of ice.

RARE EARTH ELEMENTS Chemically active metals such as yttrium, lanthanum, and lanthanides that are rare in the Earth's crust.

RECEPTACULITID A problematic Paleozoic marine sessile calcareous organism with egg-shaped skeletons built of elements arranged in whorls around the central axis.

REDBED A bed of terrigenous SEDIMENTARY ROCK that has oxidized through exposure to air, turning its iron components rusty red. Red beds are often associated with SANDSTONE.

REEF A CARBONATE deposit, formed by accumulated skeletons of CORAL, that forms an important marine habitat. Fringe reefs build up on the shores of continents or islands, the living animals mainly occupying the outer edges; barrier reefs are separated from the shore by saltwater lagoons as much as 18mi (30km) wide; atolls surround lagoons and form where an extinct volcano has subsided.

RELICT POPULATION A group of animals or plants that survive in a limited area after having been much more widespread.

REPTILE Any member of the VERTEBRATE class Reptilia, which includes snakes, turtles, alligators, and crocodiles. Reptiles evolved from AMPHIBIANS in the Carboniferous period. Some ancient forms, such as the PLESIOSAURS and ICHTHYOSAURS, lived in the sea; modern reptiles live on land. They are cold-blooded and reproduce by means of a hard-shelled egg, the device that allowed them to colonize land.

RHABDOSOME The protective casing for a complete colony of GRAPTOLITE zooids.

RHEIC OCEAN The ocean that separated AVALONIA and GONDWANA in the early Paleozoic.

RIFT An elongated depression formed by the downward movement of an area of land between parallel systems of faults. Rifts occur in regions of crustal extension, where lithospheric plates are diverging and a continent is breaking apart. Rifts tend to form valleys.

RING OF FIRE The seismically active borders of the Pacific Ocean, shown by the frequency of earthquakes and the abundance of volcanoes, caused by the Benioff zones produced by the subduction of the Pacific plate, the Cocos plate, and the Nazca plate.

RNA Ribonucleic acid, a nucleic acid present in all cells. Several different type of RNA play a part in the mechanisms by which DNA directs the synthesis of PROTEINS in a cell.

ROCHE MOUTONNEE An exposed rock that has been polished on one side and pulled apart on the other by the passage of a GLACIER over it.

RODINIA The Precambrian supercontinent that consisted of parts of all the modern continents except Africa.

RUGOSAN An extinct solitary or branching horn-shaped modular CORAL from the Ordovician and Permian.

RUMINANT An animal that chews the cud, such as a cow.

S

SALT A compound consisting of a metal and a base, as is formed when an acid has its hydrogen replaced by a metal. Common salt, NaCl, is the sodium salt of hydrochloric acid.

SALT DOME A DIAPIR formed from salt. As a bed of rock salt is compressed it deforms plastically and rises through the overlying beds, twisting them upwards.

SAN ANDREAS FAULT A TRANSFORM FAULT along the coast of California that gives rise to the many earthquakes in the area. Its continued movement will probably cause that part of California to shear off and drift away over the next few million years.

SARCOPTERYGIAN See LOBE-FINNED FISH.

SAURISCHIAN A "lizard-hipped" dinosaur, from which (in spite of the name) all birds descended. Saurischians included both meat-eaters and long-necked plant-eaters.

SAUROPOD A member of a group of plant-eating SAURISCHIAN dinosaurs characterized by their long necks.

SAVANNAH A landscape of tropical grasslands with scattered trees, typical of the area between the Earth's equatorial rainforests and the tropical desert belt.

SCAVENGER An animal that feeds on the flesh of dead animals killed by other animals.

SCHIST A METAMORPHIC ROCK (such as mica) that tends to be split into layers by increased temperature and pressure, causing the FELSIC and MAFIC constituents to separate. This gives metamorphic rock a banded appearance with alternate layers of felsic and mafic crystals.

SCLERACTINIAN A member of the order (Scleractinia) to which most corals have belonged since the Paleozoic. It includes modern corals.

SCLERITE A scalelike or spinelike hollow element of a skeletal cover (SCLERITOME).

SCLERITOME A skeletal cover consisting of isolated SCLERITES.

SEAFLOOR SPREADING The process by which the ocean floor grows as new crust emerges and moves laterally away from MID-OCEAN RIDGES. Observation of this in the 1960s, combined with the theory of CONTINENTAL DRIFT, generated the idea of PLATE TECTONICS.

SEAMOUNT An isolated submarine uplift more than 3250ft (1000m) high.

SEDIMENTARY ROCK A rock formed by the accumulation and solidification of layers of fragments to form a solid mass. There are three types: CLASTIC, such as SANDSTONE, in which the fragments are derived from the breakdown of pre-existing rocks; BIOGENIC, such as COAL, in which the fragments are derived from once-living things; and chemical, such as rock salt, which is formed from crystals precipitated from solution in water.

SEED FERNS An extinct group of varied Carboniferous plants that reproduced by means of seeds rather than spores. However, they were not technically "ferns" at all, just fernlike in appearance.

SEISMIC WAVES Vibrations sent out by an earthquake. They take a number of forms, including P-waves, which are waves of compression; S-waves, which cause the shaking action; and L-waves that travel on the surface and cause the damage.

SEISMOLOGY The study of earthquakes and of the passage of vibrations through the Earth.

SEPTUM A wall of bone or shell that divides a hollow in a skeleton into separate chambers.

SESSILE Describing an immobile organism that lives on the sea floor.

SEVIER OROGENY An episode of IGNEOUS and FOLD-AND-THRUST activity north of California during the Cretaceous period.

SERIES A stratigraphic unit consisting of rocks deposited or emplaced during an epoch.

SHALE A fine-grained SEDIMENTARY ROCK, formed from the solidification of mud, easily split into fine layers or flakes.

SHELL BED A CARBONATE or phosphate bed consisting of fossil shells.

SHELF A continental margin that is covered with water, forming shallow seas.

SHELF SEA A sea that covers a continental shelf and is much shallower than true ocean. The North Sea is an example.

SHIELD Another term for a continental CRATON.

SHOCKED QUARTZ A mineral such as QUARTZ or FELDSPAR with closely-spaced microscopic layers within it, caused by the intensely high pressures of an impact shock, as when a meteorite strikes the Earth.

SIDEROPHILE ELEMENT A chemical element that has an affinity for the metal phase; for example, iron or nickel. During the Earth's formation the siderophile elements sank towards the core.

SILICATE A MINERAL that is a compound of metallic elements with silicon and oxygen.

SILICICLASTIC A SILICATE and MINERAL sediment derived mostly from weathering land.

SILL An emplacement of IGNEOUS ROCK, intruded between beds of SEDIMENTARY ROCK.

SIPHUNCLE A canal that runs through all the chambers of a NAUTILOID or AMMONITE shell, adjusting the air pressure to affect buoyancy.

SOLAR NEBULA The cloud of dust and gas from which the solar system eventually condensed after the BIG BANG.

SONOMA OROGENY A mountain-building event at the Permian–Triassic boundary as an eastward-moving ISLAND ARC collided with the Pacific margin of North America.

SPECIATION The process by which new SPECIES appear and change over time.

SPECIES The basic level of taxonomic classification; a group of organisms that can interbreed and produce fertile offspring. Related species are grouped together in a genus.

SPHENOPSID (Equisiophyta) A group of spore-bearing plants such as the giant horsetail *Calamites*, common in the Late Paleozoic.

SPICULE A tiny needle-like calcareous or siliceous structure, forming part of the skeleton of an INVERTEBRATE animal.

SPONGE A primitive sessile aquatic multicellular animal with an aquiferous system and a body enclosed by a covering tissue. Sponges appeared in the latest Proterozoic.

SPORANGIUM The structure on a plant that holds the SPORES.

SPORE A reproductive plant body that consists mostly of a cell with half the viable number of CHROMOSOMES. It must unite with another spore before growing into a plant.

STABLE ZONE A region of the Earth's crust that is not subject to OROGENY or other deformational processes. Stable zones are typically found within continental interiors, away from margins and MOBILE BELTS.

STAGE A stratigraphic unit smaller than a series or epoch.

STRATA Layers or beds of SEDIMENTARY ROCK.

STRATIGRAPHY The study of the relationships, classification, age, and correlation of rock STRATA that lie at or near to the surface of the planet. The succession of these rocks enables scientists to build up a geological history of the Earth.

STRIKE-SLIP FAULT A FAULT in which one body of rock moves sideways rather than vertically in relation to the next.

STROMATOLITE A laminated structure formed in quiet water when a layer of filamentous ALGAE traps sedimentary particles, chiefly CARBONATE. Another layer of algae grows on this sedimentary surface, trapping another layer, so building up a dome shape or a column. Fossil stromatolites are known from Precambrian times, when there were no other life forms to disturb their growth.

STROMATOPOROID One of a group of extinct calcified marine SPONGES that built REEFS.

SUBAERIAL Taking place on land. REDBEDS are subaerial SEDIMENTARY ROCKS.

SUBDUCTION The movement of one LITHO-SPHERIC plate as it slides beneath another into the MANTLE and is consumed. The process is an integral part of PLATE TECTONICS.

SUBDUCTION ZONE The inclined region where subduction of the LITHOSPHERE occurs.

SUPERCONTINENT A continent made up from the amalgamation of more than one continental mass.

SUPERNOVA The explosion that results when the structure of a star collapses.

SYMBIONT An organism that coexists with and depends on another organism.

SYNAPSID A member of one of the major subclasses of REPTILES, including mammal-like reptiles and MAMMALS. They had a characteristic skull with an extra opening on each side. *See* DIAPSID.

SYSTEM (geological) A stratigraphic unit consisting of all the rocks deposited or emplaced during a geological period.

T

TABULATE A form of CORAL that existed from the early to the late Paleozoic.

TACONIC OROGENY An early phase of the APPALACHIAN OROGENY, which took place during the Ordovician during the accretion of ISLAND ARCS to Laurentia.

TAENIODONT A member of a group of primitive MAMMALS from the earliest Tertiary.

TARDIGRADE A microscopic INVERTEBRATE that evolved in the Cambrian. It possessed four segments covered with a firm cuticle and bearing a pair of telescopic legs each.

TARDIPOLYPOD A Cambrian marine wormlike INVERTEBRATE with a segmented body and multiple telescopic legs.

TAXONOMY The classification of organisms into groups (taxa). The basic unit is the SPECIES, progressing up through genus, family, order, class, phylum, and kingdom.

TECTONIC PLATE A section of LITHOSPHERE that moves as a unit during PLATE TECTONICS. Usually a plate grows along one edge, at a MID-OCEAN RIDGE, and is destroyed along another, at an OCEAN TRENCH.

TECTONOEUSTASY Global sea level fluctuation caused by absolute changes in the quantity of seawater as MID-OCEAN RIDGES grow.

TELEOST A group of fish with a bony skeleton, small rounded scales, and a symmetrical tail. Most modern fish are teleosts.

TERRANE A relatively small block of the Earth's crust that is distinct from those around it.

TERRIGENOUS Relating to rocks or sediment that have been formed of material eroded from landmasses.

TERTIARY The era of geological time between the Mesozoic and the Quaternary. It encompasses the last 65 million years of Earth's history except for the last two million or so.

TETHYS SEAWAY (or Ocean) An oceanic region that existed as a vast embayment into PANGEA, almost separating LAURASIA from GONDWANA. It was lost as Africa and India closed with Europe and Asia, leaving behind the Mediterranean, Black, Caspian, and Aral Seas.

TETRACORAL *See* RUGOSAN.

TETRAPOD Any VERTEBRATE that is not a fish. Although the name means "four-footed," the classification also covers whales, birds, and snakes, whose ancestors were real tetrapods with Devonian origins.

THALLOPHYTE A primitive plant, such as a seaweed, in which the body (thallus) is not divided into roots or stems or leaves, or any of the other features associated with more advanced plants.

THECODONT A crocodile-like Triassic REPTILE ancestral to the dinosaurs.

THERAPSID A member of a group of advanced mammal-like REPTILES.

THEROPOD A member a group of meat-eating SAURISCHIAN dinosaurs.

THRUST FAULT A low-angle fault produced by compression. In mountain-building, thrust sheets (large slices of rock) can slide horizontally over underlying rocks for great distances.

TILL An unsorted mixture of clay and cobbles deposited by GLACIERS in MORAINE.

TILLITE A rock formed by the lithification of TILL.

TILLODONT A member of a group of primitive plant-eating MAMMALS from the early Tertiary, probably closely related to the TAENIODONTS.

TITANOTHERE See BRONTOTHERE.

TOMMOTIID A problematic marine INVERTEBRATE of the Cambrian that was covered with phosphatized ribbed SCLERITES.

TORNQUIST SEA A sea that occupied the western part of BALTICA in the early Paleozoic.

TRACE FOSSIL A fossilized trail, track, borehole, burrow, or footprint.

TRACHEOPHYTE A multicellular land plant with distinct tissues and organs and developed vascular system.

TRADE WINDS The prevailing winds that blow towards the equator, caused by hot tropical air rising and drawing in the cooler air from north and south. They blow from the southeast and the northeast, deflected in these directions by the Coriolis effect of the Earth's rotation.

TRANSFORM FAULT A geological fault produced at a MID-OCEAN RIDGE and formed as adjacent TECTONIC PLATES slide past one another.

TRANSGRESSION The gradual encroachment of sea over a land area.

TRAP A stairlike structure formed from successive basaltic LAVA flows over a wide area, as seen in the Deccan and Siberian traps.

TRILOBITE A Paleozoic marine ARTHROPOD that scavenged on the bottom of the shallow seas. Trilobites had plated, segmented bodies with many Y-shaped legs—they resembled modern wood lice. They died out at the end of the Permian but are abundant as fossils in Paleozoic rocks.

TRIPLE JUNCTION A point at which three LITHOSPHERIC plates meet. Ocean ridges often meet at triple junctions, which accounts for the often jagged nature of continental margins.

TROPHIC WEB A system of continuous chains of SPECIES where each link is a species consumable by subsequent species; this web transforms the energy in an ECOSYSTEM.

TUNDRA A landscape of stunted seasonal vegetation, snow-covered in winter and flooded in summer, caused by PERMAFROST and typical of far northern continental areas.

TURBIDITY CURRENT Moving water that contains suspended sediment, making it denser than surrounding water, and causing it to flow at a lower depth along the seafloor.

TYLOPOD A member of a group of even-toed UNGULATES that includes the camels.

U

ULTRAMAFIC CRUST Heavy crust, like MAFIC crust but containing even less silica.

UNCONFORMITY A break in the sequence of deposition of SEDIMENTARY ROCKS, formed when a sequence of rocks is raised above sea level and eroded, and then submerged so that deposition resumes.

UNGULATE Any four-legged hoofed mammal.

UNICELLULAR An organism whose entire body consists of a single cell.

UNIFORMITARIANISM The principle that the natural laws and processes that form present-day rocks and landscapes have remained the same over time, so that ancient geological formations and their processes may be interpreted by observing analogous formations and processes in the world today. This is expressed as "The present is the key to the past." However, the rate at which processes operate may have been different in the distant past, and their relative importance may have changed also.

UNIRAMIAN An ARTHROPOD whose appendages are not branched.

UPWELLING (marine) The movement of deep water, usually off the coast of a continent, that brings nutrients closer to the surface where PLANKTON and other organisms feed.

URALIAN OCEAN The ocean separating Siberia and BALTICA in the early Paleozoic.

UROCHORDATE A sea squirt—a mostly sessile, sacklike marine CHORDATE lacking both chord and notochord when adult. Sea squirts appeared first in the Carboniferous.

U-SHAPED VALLEY A valley that has been ground downwards and sideways by the weight of a GLACIER so that it has a flat bottom and vertical sides.

V

VARIATION A difference between individuals of the same SPECIES, found in any sexually reproducing population, due to genetic or environmental factors or their combination.

VARVE A thin layer of sediment deposited in a glacial lake. GLACIERS melt at different rates between seasons, and the meltwater carries away a different load of TILL. Varves build up as annual cycles of coarse and fine material, which can be used by geologists to investigate a glaciated area.

VASCULAR PLANT A plant with a plumbing system that can carry food and water around its body. All plants more advanced than mosses are vascular plants.

VERTEBRATE Any animal that has a backbone. There are approximately 41,000 vertebrate species, including mammals, birds, reptiles, amphibians, and fish.

VESTIMENTIFERAN A wormlike marine INVERTEBRATE that appeared in the Silurian.

VIVERRAVINE The cat branch of the CARNIVORES: a group of primitive meat-eating MAMMALS that evolved from MIACIDS in the early Tertiary, and the line from which hyenas, mongooses, civets, and all felids (cats) evolved.

VOLATILE ELEMENT Any of the elements that show an affinity for the ATMOSPHERE, including hydrogen, nitrogen, carbon, oxygen, and the inert gases (helium, argon, neon, krypton, and xenon).

VOLCANIC ARC See ISLAND ARC.

VULPAVINE The dog branch of the CARNIVORES: a group of primitive meat-eating MAMMALS that evolved from MIACIDS in the early Tertiary, and diversified into bears, foxes, wolves, mustelids, seals and sea lions, pandas, and all true dogs (canids).

W

WALLACE'S LINE The boundary between the BIOGEOGRAPHIC PROVINCES of Australia and southeast Asia, which runs through the strait between Bali and Lombok.

WATER COLUMN A vertical section through the sea or a lake, highlighting the differences in properties of the water at different levels.

WEATHERING The chemical or physical processes by which exposed rock is broken down by rain, frost, wind, and other elements of the weather. It is the beginning of EROSION.

WILLISTON'S LAW The evolutionary principle which states that serially arranged structures in animals, such as teeth and legs, will become fewer and take on new functions as new species evolve. For example, MAMMALS have fewer ribs than do their fish ancestors.

WIWAXIID An extinct COELOSCLERITOPHORAN.

XYZ

XENARTHRAN A member of an order of MAMMALS that includes the armadillos, the anteaters, and the sloths.

ZEUGLODONT Having teeth shaped like arches, as in early whales.

ZONE The shortest time unit used in geology.

ZONE FOSSIL See INDEX FOSSIL.

ZOOGEOGRAPHY The study of the distribution of animal life, the animal assemblages of particular areas, and the barriers between distinct BIOGEOGRAPHIC REALMS.

ZOOID An individual, clonally produced unit of a modular animal.

ZOOPLANKTON The animal component of PLANKTON: chiefly PROTOZOANS, small CRUSTACEANS, and the larval stages of MOLLUSKS and other INVERTEBRATES.

FURTHER READING

VOLUME 1

Cairns-Smith, A.G. *Seven Clues to the Origin of Life*. Cambridge, England: Cambridge University Press, 1985.

Cone, J. *Fire Under the Sea*. New York: William Morrow & Co, 1991.

Conway Morris, S. *The Crucible of Creation: The Burgess Shale and the Rise of Animals*. Oxford; New York; Melbourne: Oxford University Press. 1998.

Darwin, C. *On the Origin of Species by Natural Selection*. London: John Murray, 1859.

Decker, R. and Decker, B. *Mountains of Fire*. Cambridge, England: Cambridge University Press, 1991.

Dixon, B. *Power Unseen: How Microbes Rule the World*. New York: WH Freeman and Company, 1994.

Fortey, R. *The Hidden Landscape: A Journey into the Geological Past*. London: Pimlico, 1993.

Glaessner, M. F. *The Dawn of Animal Life*. Cambridge: Cambridge University Press, 1984.

Gould, S. J. *Wonderful Life: The Burgess Shale and the Nature of History*. New York: Norton, 1989.

Gross, M. Grant. *Oceanography: A View of the Earth*. Englewood Cliffs, NJ: Prentice-Hall, 1982.

Hsu, K.J. *Physical Principles of Sedimentology: A Readable Textbook for Beginners and Experts*. New York: Springer Verlag, 1989.

McMenamin, M. A. S. and D. L. S. McMenamin. *The Emergence of Animals. The Cambrian Breakthrough*. New York: Columbia University Press, 1990.

Margulis, L. and Schwartz, K. 1998. *Five Kingdoms: An Illustrated Guide to the Phyla of Life on Earth*. (3rd ed.) New York: WH Freeman and Company.

Norman, D. *Prehistoric Life*. London: Boxtree, 1994.

Sagan, D. and Margulis, L. *Garden of Microbial Delights: A Practical Guide to the Subdivisible World*. Dubuque, IA: Kendall-Hunt, 1993.

Schopf, J.W. *Major Events in the History of Life*. Boston: Jones and Bartlett, 1992.

Stewart, W. N. and G. W. Rothwell. *Palaeobotany and the Evolution of Plants* (2nd edition). Cambridge: Cambridge University Press, 1993.

Rodgers, J.J.W. *A History of the Earth*. Cambridge, England: Cambridge University Press, 1993.

Whittington, H. B. *The Burgess Shale*. New Haven: Yale University Press, 1985.

Wood, R. *Reef Evolution*. New York: Oxford University Press, 1999.

VOLUME 2

Alvarez, W. T. *Rex and the Crater of Doom*. Princeton, NJ: Princeton University Press, 1997.

Bakker, R.T. *The Dinosaur Heresies*. New York: William Morrow & Co, 1986.

Brusca, R.C. and Brusca, G.J. *Invertebrates*. Sunderland, Mass.: Sinauer Associates, 1990.

Currie, P.J. and Padian, K. *Encyclopedia of Dinosaurs*. San Diego: Academic Press, 1996.

Dingus, L. and Rowe, T. *The Mistaken Extinction: Dinosaur Evolution and the Origin of Birds*. New York: W.H. Freeman and Company, 1997.

Erwin, D.H . *The Great Paleozoic Crisis: Life and Death in the Permian*. New York: Columbia University Press, 1993.

Feduccia, A. *The Origin and Evolution of Birds*. New Haven: Yale University Press, 1996.

Fraser, N.C. and Sues, H–D. *In the Shadow of the Dinosaurs: Early Mesozoic Tetrapods*. Cambridge, England: Cambridge University Press, 1994.

Kenrick, P. and Crane, P. *The Origin and Early Diversification of Land Plants*. Washington, DC: Smithsonian Institution Press, 1997.

Lambert, D. *Dinosaur Data Book*. New York: Facts on File, 1988.

Lessem, D. *Dinosaur Worlds*. Hondsale, Pennsylvania: Boyd's Mill Press, 1996.

Long, J.A. *The Rise of Fishes*. Baltimore, MD and London: The Johns Hopkins University Press, 1995.

Savage, R.J.G. and Long, M.R. *Mammalian Evolution: An Illustrated Guide*. London: British Museum of Natural History, 1987.

Thomas, B.A. and Spicer, R.A. *The Evolution and Paleobiology of Land Plants*. London: Croon Helm, 1987.

VOLUME 3

Alexander, David. *Natural Disasters*. London: University College Press, 1993.

Andel, T. van. *New Views of an Old Planet*. Cambridge, England: Cambridge University Press, 1994.

Goudie, A. *Environmental Change*. London: Clarendon Press, 1992.

Hsu, K.J. *The Mediterranean Was a Desert*. Princeton, NJ: Princeton UP, 1983.

Johanson, D.C. and Edey, M.A. *Lucy: The Beginnings of Humankind*. New York: Simon and Schuster, 1981.

Lamb, H.H. *Cimate, History and the Modern World*. London: Routledge, 1995.

Lewin, R. *The Origin of Modern Humans*. New York: Scientific American Library, 1993.

McFadden, B.J. *Fossil Horses*. Cambridge, England: Cambridge Univesity Press, 1992.

Pielou, E.C. *After The Ice Age: The Return of Life to Glaciated North America*. Chicago: University of Chicago Press, 1991.

Prothero, D.R. *The Eocene-Oligocene Transition: Paradise Lost*. New York: Columbia University Press, 1994.

Stanley, S.M. *Children of the Ice Age: How a Global Catastrophe Allowed Humans to Evolve*. New York: W.H. Freeman and Company, 1998.

Tattersall, Ian. 1993. *The Human Odyssey: Four Million Years of Human Evolution*.

Tudge, C. *The Variety of Life: A survey and a celebration of all the creatures that have ever Lived*. Oxford, England: Oxford University Press, 2000.

Young, J.Z. *The Life of Vertebrates* (2nd ed.) Oxford, England: Oxford University Press, 1962.

ACKNOWLEDGMENTS

AL   Ardea London
BCC   Bruce Coleman Collection
C   Corbis
NHM   Natural History Museum, London
NHPA   Natural History Photographic Agency
OSF   www.osf.uk.com
PEP   Planet Earth Pictures
SPL   Science Photo Library

VOLUME 1

2 © Kevin Schafer/C; **3** Andrey Zhuravlev; **4** Image Quest 3-D/NHPA; **10–11** & **12-13** Royal Observatory, Edinburgh/AATB/SPL; **16** NASA/SPL; **18t** Bernhard Edmaier/SPL; **20-21** © NASA/Roger Ressmeyer/C; **23** Dr. Ken Macdonald/SPL; **26–27** © Buddy Mays/C; **28** SPL; **30** Sinclair Stammers/SPL; **32** E.A. Janes/NHPA; **33** M.I. Walker/NHPA; **35** © W. Perry Conway/C; **36** CNRI/SPL; **37** Volker Steger/SPL; **38** © Stuart Westmorland/C; **39** © Manuel Bellver/C; **40** Bruce Coleman Inc.; **42** Manfred Kage/SPL; **45** © C; **47t** A.N.T./NHPA; **47b** RADARSAT International Inc.; **48** © Kevin Schafer/C; **50–51** © Ralph White/C; **51** SPL; **54** Sinclair Stammers/SPL; **55** Martin Bond/SPL; **59** © James L. Amos/C; **60** Image Quest 3-D/NHPA; **61** © Kevin Schafer/C; **65** © Stuart Westmorland/C; **66–67** & **68–69** Paul Kay/OSF; **71** Andrey Zhuravlev; **74** P.D. Kruse; **77** Digital image © 1996 C: Original image courtesy of NASA/C; **78** Andrey Zhuravlev; **80t** S. Conway Morris, University of Cambridge; **81** © Raymond Gehman/C; **85** Andrew Syred/SPL; **87** © Raymond Gehman/C; **90t** © David Muench/C; **90b** Breck P. Kent/OSF; **92–93** Rick Price/Survival Anglia/OSF; **94** & **96** Sinclair Stammers/SPL; **99t** P.D. Kruse; **99c** Andrey Zhuravlev; **102** © James L. Amos/C; **106** Laurie Campbell/NHPA; **108** © Ralph White/C; **110** Jens Rydell/BCC; **112** Sinclair Stammers/SPL; **115** Breck P. Kent/Animals Animals/OSF; **116** Sinclair Stammers/SPL; **120** Norbert Wu/NHPA.

VOLUME 2

2 © Scott T. Smith/C; **3** © James L. Amos/C; **4** Richard Packwood/OSF; **10–11** & **12–13** Alfred Pasieka/SPL; **18** Jane Gifford/NHPA; **20** Jon Wilson/SPL; **20–21** © Jonathan Blair/C; **23t** NHM; **23b** & **27** © James L. Amos/C; **32** Oxford University Museum of Natural History; **32–33** © Patrick Ward/C; **35** Trustees of The National Museums of Scotland; **37** Richard Packwood/OSF; **43** © David Muench/C; **45** Tony Craddock/SPL; **50** George Bernard/SPL; **56** Tony Waltham/Geophotos; **57b** NHM; **58** Brenda Kirkland George, University of Texas at Austin; **59** © Buddy Mays/C; **60** Hjalmar R. Bardarson/OSF; **65** © Jonathan Blair/C; **66–67** & **68–69** François Gohier/AL; **75t** © Scott T. Smith/C; **75b** NHM; **76** © David Muench/C; **77t** Jane Burton/BCC; **77b** © Kevin Schafer/C; **81** C. Munoz-Yague/Eurelios/SPL; **87t** © C; **88** Jane Burton/BCC; **89** NHM; **90** François Gohier/AL; **92** © James L. Amos/C; **98** Ken Lucas/PEP; **100** © Michael S. Yamashita/C; **104** Ron Lilley/BCC; **106–107** U.S. Geological Survey/SPL; **108** Martin Bond/SPL; **111** & **112** François Gohier/AL; **112–113** Louie Psihoyos/Colorific; **116–117** © C; **119** SPL.

VOLUME 3

2 © Michael S. Yamashita/C; **3** NHM; **4** Anup Shah/PEP; **10–11** & **12–13** Jeff Foott/BCC; **16** John Mason/AL; **18** Digital image © 1996 C: Original image courtesy of NASA/C; **20** Patrick Fagot/NHPA; **22** © Douglas Peebles/C; **24–25** Dr. Eckart Pott/BCC; **26** Tony Waltham/Geophotos; **27** NHM; **28t** S. Roberts/AL; **29t** John Sibbick; **29b** NHM; **30** Bruce Coleman Inc.; **32** © Jonathan Blair/C; **36** AL; **39** Anup Shah/PEP; **42** CNES, 1986 Distribution Spot Image/SPL; **46–47** © Michael S. Yamashita/C; **48–49** © Liz Hymans/C; **49** Digital image © 1996 C: Original image courtesy of NASA/C; **50** François Gohier/AL; **51** B & C Alexander/PEP; **53b** BCC; **56** Ferrero–Labat/AL; **56–57** NHM; **58–59** © Sally A. Morgan; Ecoscene/C; **59t** & **59b** NHM; **63** G.I. Bernard/NHPA; **64** Nigel J. Dennis/NHPA; **65** Andy Rouse/NHPA; **66–67** & **68–69** F. Jalain/Robert Harding Picture Library; **74** Peter Steyn/AL; **76–77** Simon Fraser/SPL; **78** Wardene Weisser/AL; **79** David Woodfall/NHPA; **80–81** M. Moisnard/Explorer; **82** François Gohier/AL; **84** Kevin Schafer/NHPA; **85** NHM; **86l** NASA/SPL; **86r** inset Jane Gifford/NHPA; **87** Chris Collins, Sedgwick Museum, University of Cambridge; **88l** Volker Steger/Nordstar-4 Million Years of Man/SPL; **89** NHM; **90** J.M. Adovasio/Mercyhurst Archaeological Institute; **91** © Gianni Dagli Orti/C; **94** © Peter Johnson/C; **97** Sheila Terry/SPL; **98** inset © Charles & Josette Lenars/C; **100** NASA/SPL; **102** Matthew Wright/Been There Done That Photo Library; **104–105** © Galen Rowell/C; **106–107** Tom Bean; **107** Luiz Claudio Marigo/BCC; **108–109** A.N.T./NHPA; **109** Felix Labhardt/BCC; **110t** © Mike Zens/C; **110b** Adrian Warren/AL; **111t** © Robert Pickett/C; **111b** © Eric Crichton/C; **112c** Steven C. Kaufman/BCC; **112b** © Clem Haagner; Gallo Images/C; **112–113** Gunter Ziesler/BCC; **115** Jeff Foott/BCC; **116** Erich Lessing/Archiv für Kunst und Geschichte; **116–117** © Yann Arthus-Bertrand/C; **118** David Woodfall/NHPA; **118–119** D. Parer & E. Parer-Cook/AL; **120** Mark Conlin/PEP.

GENERAL ACKNOWLEDGMENTS

We would like to thank Dr. Robin Allaby of the University of Manchester Institute of Science and Technology (UMIST) and Dr. Angela Milner of the Natural History Museum, London for their specialist help, and John Clark, Neil Curtis, and Sarah Hudson for editorial assistance.

INDEX